CW00953982

VEGETA

TAS

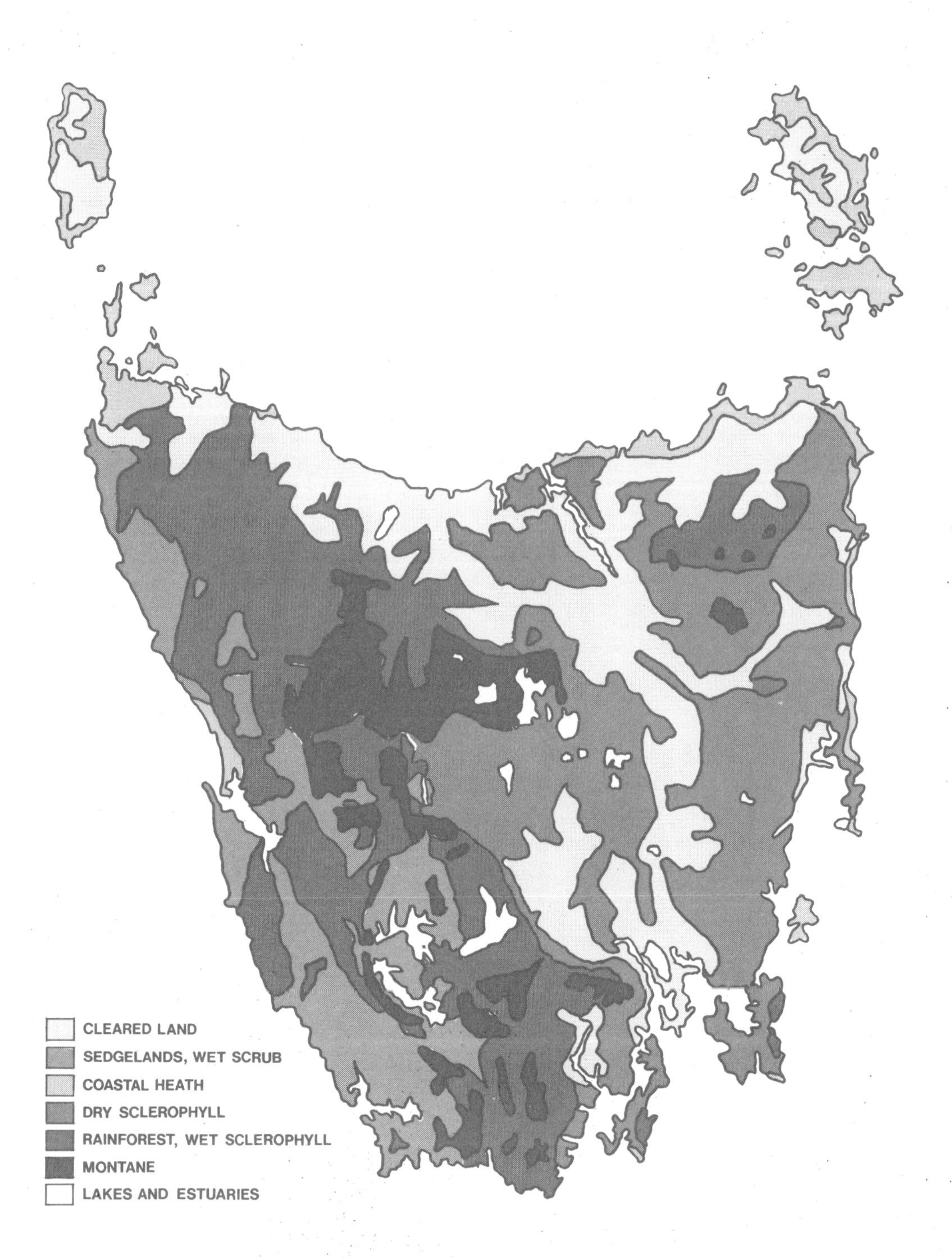
CLEARED LAND
SEDGELANDS, WET SCRUB
COASTAL HEATH
DRY SCLEROPHYLL
RAINFOREST, WET SCLEROPHYLL
MONTANE
LAKES AND ESTUARIES

GUIDE TO FLOWERS AND PLANTS OF TASMANIA

GUIDE TO FLOWERS AND PLANTS OF TASMANIA

LAUNCESTON FIELD NATURALISTS CLUB

Edited by Mary Cameron, B.Sc., A.L.A.A.

REED

First published 1981
Reprinted 1984
Reprinted 1986
Reprinted 1992
Reprinted 1994

Reed Books
a division of
Reed International Books Australia Pty Ltd
Level 9, North Tower
Chatswood Plaza, 1-5 Railway St,
Chatswood, NSW 2067

National Library of Australia
Cataloguing-in-Publication data
Guide to flowers and plants of Tasmania.

At head of title: Launceston Field
Naturalists Club.
Includes index.
ISBN 0 7301 0169 X.

1. Botany – Tasmania. 2. Plants – identification.
I. Cameron, Mary. II. Launceston Field Naturalists Club.

581.19946

Printed and bound in Singapore

CONTENTS

FOREWORD

This is a book for all those who enjoy Tasmania's heritage of flowering plants. In *Flora Tasmaniae* (1860), the first comprehensive book on Australian flora to be published, J. D. Hooker wrote: 'The Flora of Australia has been justly regarded as the most remarkable that is known, owing to the number of peculiar forms of vegetation which that continent presents'. When the plants were first seen in Europe they caused excitement and amazement with their novelty: and immense problems in classification and nomenclature.

A glimpse of Australian vegetation was seen in England in 1688 and 1699 when a few specimens collected by William Dampier from the north-western shores of the continent were received. But the country's first botanical investigators were the eighteenth century naturalists of Captain James Cook's three voyages of exploration in the Pacific Ocean. During the third voyage, in 1777, Cook's two ships, the *Resolution* and *Discovery*, anchored in Adventure Bay, Bruny Island, for four days. Among the plants collected by William Anderson and David Nelson and subsequently taken to England was the one named by l'Héritier *Eucalyptus obliqua,* the type species of this new genus.

The southern lands also lured explorers from France; an expedition that reached southern Tasmania in 1792 under the command of Admiral Bruny d'Entrecasteaux included the naturalists Labillardière and Riche: in our flora genera and species commemorate their names.

The years at the beginning of the nineteenth century were years of difficulty for the early settlers in Tasmania but a time of great botanical achievement. In 1801 Captain Matthew Flinders began his voyage to survey the coasts of Terra Australis; the naturalist of the expedition was Robert Brown. Brown's appointment was due to Sir Joseph Banks who had sailed with Cook's first voyage (1768-1771) and whose discoveries laid the foundations of knowledge of the Australian flora. Brown made extensive collections along the south and east coasts of mainland Australia and then set out for 'Van Diemen Island'. He was storm bound on islands of the Kent Group in Bass Strait but reached Port Dalrymple in January 1804. Then, by sailing to Port Phillip, he joined Lieut.-Governor David Collins's first group of settlers to the River Derwent at Risdon Cove and later at Hobart. Here, because of lack of shipping, he stayed for nine months and collected seven hundred species of flowering plants to add to his collections from the mainland. On returning to England he published, in 1810, *Prodromus Florae Novae Hollandiae et Insulae Van Diemen.* This book marks a new epoch in botanical thought, for Brown not only described his plants, he classified them in a natural system which took account of a number of characters and led to grouping of plants according to their relationships.

Robert Brown's work attracted the special interest of the Professor of Botany at the University of Glasgow (the later Director of The Royal Botanic Gardens, Kew), Dr William Jackson Hooker, who made great efforts to obtain more specimens from Van Diemen's Land. Through a former resident of Glasgow who had settled in Launceston, Thomas Scott, he found, in 1830, a correspondent in Robert Lawrence, son of W. E. Lawrence, a landowner in the north of the island. Robert Lawrence became an enthusiastic collector and student of plants and excited the same interests in his friend Robert Campbell Gunn.

After the untimely death of Robert Lawrence in 1833 Gunn continued to develop his work. In 1840 he met the son of W. J. Hooker, Joseph Dalton Hooker, who spent some months in Tasmania during his appointment as assistant surgeon and botanist of the Antarctic expedition of 1839-1843; the ships *Erebus* and *Terror* were under the command of Captain Sir James Clark Ross. Hooker in the introduction to his classic *Flora Tasmaniae* wrote: 'I can recall no happier weeks of my various wanderings over the globe than those spent with Mr Gunn, collecting in the Tasmanian mountains and forests, or studying plants in his library, with the works of our predecessors Labillardiere and Brown'. He also pays tribute to the generous help of his friend William Archer whose estate, 'Cheshunt', was on the Meander River near the foot of the Western Tiers.

The co-operation which began between Robert Lawrence and W. J. Hooker extends to the present through a succession of naturalists sharing their observations, studies and knowledge of our plants. The members of the Launceston Field Naturalists Club must be congratulated on maintaining a great tradition.

Winifred M. Curtis, A.M., D.Sc.(Lond.), F.L.S.

ACKNOWLEDGMENTS

I should like to thank members of the Launceston Field Naturalists Club who have contributed in many ways, and all those who have been involved directly with the preparation of the book — in researching, writing, typing, checking descriptions, selecting slides or preparing art work; in particular, Wyn Atkins, Evelyn Bach, Eleanor and Geoff Best, Ann and Cec Bird, Maren Bjerring, Dulcie and Terence Butler, Deborah and Rosemary Cameron, Marjorie and Humphrey Gardner, Betty Gee, Dianne Hoggins, Sheridan and Eric Hrycyszyn, Ada Plumridge, Marion Simmons, Jean and Bill Stephens and Ruth Upson.

However, it has been the capable direction and dedication of our editor, Mary Cameron, that has guided the work to its conclusion.

Many photographers have been most generous in allowing the use of their colour slides. They have been acknowledged in the credits, but I should like to thank especially Betty Gee, Lucy King and Ada Plumridge for making available the slide collections of their late husbands. The H. J. King collection has helped to fill many gaps and the extent of our use is obvious from the credits.

Finally, I should like to thank our Patron, Dr Winifred Curtis, for her encouragement and assistance, and for writing the foreword.

John Simmons
Vice President
Launceston Field Naturalists Club

INTRODUCTION

Members of the Launceston Field Naturalists Club have been asked repeatedly to recommend a simple illustrated non-technical book for the identification of wild-flowers in Tasmania. It is to fill this need that *A Guide to Flowers and Plants of Tasmania* has been produced, for use by ordinary nature lovers, tourists, bushwalkers and students. The idea of writing such a book using some of the many excellent slides taken by present and past members was first put forward by Club President, John Simmons, under whose guidance the work has proceeded.

It is impossible to feature all the flowering plants — nearly 2000 species are recorded for Tasmania — so a representative sample of those easily seen has been chosen. Many are common and widespread, others are significant parts of the Tasmanian flora, either confined to Tasmania or examples of genera more richly represented in Tasmania than in other States.

Although some are less common than others there is no plant shown which cannot be seen within a half-day's walk from a road somewhere in Tasmania and without using a four-wheel-drive vehicle.

For convenience in the use of this book, the vegetation has been divided into several types — the temperate rainforest or myrtle forest, wet sclerophyll or eucalypt forest, open or dry sclerophyll forest, montane and subalpine and coastal heath and coastal light forest. Some plants are widespread, growing from sea level to mountain tops, others are common on river banks and wet places; there are sections on these also. Notes on distribution are given with each description, and there is fuller discussion of vegetation in the botanical introduction.

INTRODUCTION TO THE THIRD PRINTING

We are indebted to the many friends who have pointed out error in spelling and text since our first volume was published.

In this printing we are taking the opportunity to correct a more serious error. Dr. R. W. Barker of the State Herbarium, Botanic Gardens, Adelaide has identified the Euphrasia plates as follows:

Plate 35 Euphrasia collina subspecies diemenica not E. gibbsiae
Plate 47 E. collina ssp diemenica, a name change only
Plate 52 E. gibbsiae ssp comberi, not E. striata

Textual changes have been made in accordance with his identifications. We thank Dr. Barker very much for his assistance in this difficult genus.

INTRODUCTION TO THE FOURTH PRINTING

Since the third printing of this book in 1986, there has been an upsurge in botanical research in an effort to document the Australian Flora and to standardise the nomenclature. Consequently some names used in this book differ from those in previous printings but they refer to the same plants.

To bring this edition up to date it has been necessary to refer to the following publications:

A. M. Buchanan, A. McGeary Brown and A. E. Orchard, *A Census of the Vascular Plants of Tasmania.*

Tasmanian Herbarium Occasional Publication No. 2, 1989, D. I. Morris, *The Grasses of Tasmania.*

Tasmanian Herbarium Occasional Publication No. 3, 1991, Bureau of Flora and Fauna, Canberra, relevant volumes of *Flora of Australia,* AGPS.

Mary Cameron

HOW TO USE THIS BOOK

Each photograph is accompanied by a description which includes the average size of the plant, its flowering time and the kind of locality you may expect to find it. The botanical names have two parts, the first word is the name of the *genus* and the second that of the *species*. Common names are given where they are in common use, and the *family* to which the plant belongs. The descriptions are not full but usually mention the important criteria for identification.

To identify a specimen a careful comparison should be made with the photograph and the description. Plants are identified primarily on the structure of the flower, the form and arrangement of leaves and on the degree of woodiness, that is, whether tree, shrub or herb. Therefore in using this book to identify a particular plant it is necessary to look carefully at the structure of the flower. Details to observe include the way the flowers are arranged, their shape, the number and arrangement of sepals, petals and stamens, whether they are free or joined to one another, the position and shape of the ovary and fruit.

In a few descriptions where the family is small or there are only two or three different species in Tasmania, only one species is illustrated but the description mentions the other species and gives diagnostic features for their identification. This is not possible in larger genera. Where two plants with the same generic name are mentioned in a description the second name is given with an initial only, not the name in full — thus '*Tetratheca pilosa* is larger than *T. procumbens*' means '. . .than *Tetratheca procumbens*'.

If the plant you wish to identify is not illustrated, compare the flower structure with the illustrations. If there is agreement between flowers but not leaves, the identification will probably be correct to the level of family, perhaps of genus. It will then be necessary to use a more complete work.

Full botanical descriptions and keys for identification will be found in W. M. Curtis, *The Student's Flora of Tasmania,* Parts 1, 2, 3 and 4A. A simple hand lens giving X 10 magnification is useful. Descriptions in this book do not use details too small to be seen with such an aid.

There are few grasses, sedges or rushes featured because the identification of these Monocotyledons involves the dissection under the microscope of their very small flowers, and the terms used to express the differences are very technical, involving too much explanation for this work. Part 4B of *The Student's Flora of Tasmania* will deal with such plants but is still in progress.

ABBREVIATIONS

NSW	New South Wales
NT	Northern Territory
Qld	Queensland
SA	South Australia
Tas	Tasmania
Vic	Victoria
WA	Western Australia
sp., spp.	species (sing. and plural)
var.	variety
±	more or less

PHOTOGRAPHIC CREDITS

Geoff Best: 3, 5, 6, 9, 14, 35, 81, 84, 91, 96, 100, 105, 108, 115, 137, 138, 139, 150, 152, 172, 188, 194, 208, 221, 227, 230, 232, 234, 242, 248, 249, 251, 270, 274, 276.
Athol Beswick: 10, 25, 78, 107, 118, 123, 187, 189, 191, 206, 216, 286.
Maren Bjerring: 18, 45, 66, 106, 121, 127, 144, 168, 169, 184, 209, 224, 236, 239, 250, 265, 268, 278.
Douglas Bryan: 143, 252, 280, 293.
Terence Butler: 92, 97, 120, 149, 155, 178, 190, 212, 238, 240, 256, 285, 299.
Jean Carins: 62, 218.
Nigel Davies: 42, 52.
Penny Evans: 11, 26, 38.
The late Mary Fisher: 131, 133, 182, 258.
Berry Fowler: 15, 31, 55, 60, 70, 95, 103, 104, 146, 154, 193.
The late J. N. Gee: 1, 2, 50, 59, 61, 145, 166, 180, 199, 203, 214, 219, 229, 231, 259, 287, 291.
Jim Glennie: 8, 28, 30, 46, 73, 173, 197, 246.
D. Glenny: 43.
Margaret Green: 13, 22, 44, 65, 86, 114, 117, 126, 237, 279.
Eric Hrycyszyn: 49, 102, 156, 171, 176, 192, 228, 235.
The late H. J. King: 7, 24, 33, 36, 39, 51, 53, 56, 57, 69, 72, 74, 75, 79, 80, 85, 94, 111, 112, 113, 122, 124, 125, 132, 136, 147, 148, 157, 158, 160, 161, 163, 164, 170, 181, 185, 186, 201, 207, 213, 222, 233, 244, 254, 264, 266, 297.
Michael Larner: 12, 67, 134, 142, 153.
The late M. R. Plumridge: 48, 63, 76, 130, 141, 284, 292, 300.
John and Marion Simmons: 4, 16, 17, 19, 20, 21, 22, 23, 27, 29, 32, 34, 37, 40, 41, 47, 54, 58, 64, 68, 71, 77, 82, 83, 87, 88, 89, 90, 93, 98, 99, 101, 109, 110, 116, 119, 128, 129, 140, 151, 159, 162, 165, 167, 174, 175, 177, 179, 183, 195, 196, 197, 198, 200, 202, 204, 205, 210, 211, 215, 217, 220, 223, 225, 226, 241, 243, 245, 247, 253, 255, 257, 260, 261, 262, 263, 267, 269, 271, 272, 273, 275, 277, 281, 282, 283, 288, 289, 290, 294, 295, 296, 298.

REFERENCES

Cochrane, G.R., Fuhrer, B.A., Rotherham, E.R., Willis, J.H., *Flowers and Plants of Victoria,* A.H. and A.W. Reed, Sydney, 1968.

Costin, A.B., Gray, M., Totterdell, C.J., Wimbush, D.J., *Kosciusko Alpine Flora,* CSIRO/Collins, Australia, 1979.

Curtis, W.M., *The Student's Flora of Tasmania,* Parts 1-4a, Government Printer, Hobart, 1963-79.

Curtis, W.M. and Morris, D.I., *The Student's Flora of Tasmania,* Part 1 (2nd ed.), Government Printer, Hobart, 1963-79.

Galbraith, J., *A Field Guide to the Wild Flowers of South-East Australia,* Collins, Sydney, 1977.

King, H.J. and Burns, T.E., *Wildflowers of Tasmania,* Jacaranda Press, Milton, Queensland, 1969.

Nicholls, W.H., *Orchids of Australia,* eds. D.L. Jones and T.B. Muir, Thomas Nelson (Aust.) Ltd, Melbourne, 1969.

Rodway, L., *Tasmanian Flora,* Government Printer, Hobart, 1903.

Stones, M. and Curtis, W.M., *The Endemic Flora of Tasmania,* Parts 1-6, Ariel Press, London, 1967-78.

Firth, M.J., *Native Orchids of Tasmania,* published by the author, Devonport, Tasmania, 1965.

Tasmanian Year Book (1979), Government Printer, Hobart.

Townrow, J.E.S., *A Species List and Keys to the Grasses in Tasmania,* Papers and Proceedings, Royal Society of Tasmania, Vol. 103, May 1969.

Willis, J.H., *Handbook to Plants in Victoria,* vols. 1 (2nd ed.) and 2, Melbourne University Press, Melbourne, 1972-78.

Barker, W. R., Taxonomic Studies in *Euphrasia.* L. (Scrophulariaceae) J. Adelaide Bot. Gard. 5, 1982

GLOSSARY

achene A small indehiscent one-seeded dry fruit, thin walled, formed from one carpel.
annual A plant which completes its life cycle in one year.
anther The part of the stamen which contains the pollen.
aromatic Odorous because of the presence of an essence or oil.
awn A stiff hair-like process from the tip of a leaf or on a grass spikelet.
axil The upper angle between the leaf and the stem which bears it.
axillary In the axil, or relating to the axil.

barbed Having small hooked hairs or bristles.
basal The lower part, or attached to the base of the stem.
beak A stout narrow projection particularly at the apex of a fruit.
berry A fleshy soft walled fruit with several hard walled seeds often separated from one another in pulp.
bilateral symmetry Of a flower which can be divided into two equal halves along one line only; such halves are mirror images.
bifid Deeply and sharply divided in two.
bipinnate Twice divided pinnately.
bract A small leaf-like structure, sometimes scale-like.
bush A small shrub without a distinct trunk.

calli (sing. *callus)* Fleshy protuberances.
calyx The outermost ring of floral parts, composed of sepals, which may be free or joined.
capsule A dry dehiscent fruit made up of several fused carpels, opening by slits or pores.
carpel A unit of the female part of the flower consisting of the ovary which contains the ovules or seeds, the stigma or receptive surface which receives the pollen and the style which joins these two parts.
caudate Produced into a tail-like appendage.
compound Composed of several parts as a compound inflorescence, where the primary axis is branched or bears lateral inflorescences.
concave Curved so that the upper surface is hollow.
cone A structure of overlapping male or female parts or flowers arranged around an axis; may be applied to flower spikes of pines, Banksia or Casuarina and to the fruits derived from these.
corolla Collective name for petals especially when joined.
creeping Spreading horizontally and rooting at the nodes.
crenate With shallow rounded teeth.
cuneate Wedge shaped; wide at the apex tapering to the stalk.
cushion plant An evergreen much branched shrub with creeping rooting branches, the terminal branches erect, with crowded leaves densely compacted, growing to an even height and forming a very firm, compact rounded cushion 1 metre or more across.
cyme An inflorescence in which the apex of the stem is terminated by a flower, and further growth is carried on by lateral shoots. Such inflorescences have the oldest flower highest up the stem axis.

dehiscent Splitting to shed the seeds.
disc florets Florets of the eye of the daisy — in this book used only where disc florets are tubular.
dorsal sepal The one at the back of the flower with its outer surface towards the stem axis.
drupe A fruit in which the ovary ripens into a soft walled fleshy fruit containing a single stone, usually with one seed.

endemic Native to the country mentioned and not occurring naturally outside it.
entire Plain, undivided, simple.
epiphytic Growing on another plant but not obtaining nourishment from it.
exserted Protruding.

falcate Sickle shaped.
filament Of a stamen — the stalk carrying the anther.
filiform Thread-like.
floret Literally a little flower, an individual flower in a complex head such as a daisy or grass.

follicle A dry fruit formed from a single carpel, opening by splitting along one side.
free Not joined to any other organ.
fruit A ripened ovary or ovaries after fertilisation, more loosely includes other persisting associated parts of the flower.

glabrous Without hairs.
glaucous Distinctly bluish green, or covered with a bluish white bloom.
globose, globular Almost spherical.

hastate Shaped like a spear head, the basal lobes triangular, spreading at only a little less than right angles to the leaf stalk.
herb A flowering plant which does not produce woody tissue.
herbaceous Of a plant, non-woody; of an organ, green and soft textured.
hermaphrodite Having both stamens and carpels.

indehiscent Not opening, of a fruit which does not open to release the seeds.
inflorescence Groups or heads of flowers, the method of arrangement differing in various ways.
internode The portion of stem between subsequent leaves or their points of origin.
irregular Having no plane of symmetry, so cannot be divided into two equal halves.

keel A sharp longitudinal ridge on the lower side, or a boat-shaped petal or petals as in a pea flower.

labellum A lip, the central petal of an orchid often very modified in form and ornamented.
leaflet A small leaf, a unit of a compound leaf.
ligulate Like a tongue, strap shaped.
lobe A projecting flap or portion of a leaf or petal.

mid-rib The central longitudinal vein of a leaf or leaf-like structure and its strengthening tissue.
montane Mountain areas below subalpine.

nectary A group of glandular cells which produce nectar.
nectar disc A disc of glandular tissue producing nectar, prominent over or surrounding the ovary.
node The point on a stem at which a leaf arises.
nut A dry indehiscent one-seeded fruit in which the ovary wall becomes tough and woody.

operculum A cap which separates along a circular line as in Eucalyptus buds or Richea flowers.
opposite Two leaves arising at the same level on opposite sides of the stem.
ovary The part of the carpel or carpels which contains the ovules or seeds.
ovoid A solid shape — oval or ovate in longitudinal section.
ovule An unfertilised seed.

panicle A compound inflorescence in which the main axis bears lateral inflorescences.
papillose With very small cylindrical or spherical fleshy protuberances.
pappus The ring of hairs or scales which make up the parachute on daisy seeds.
pappus bristle One hair of such a structure.
parasite A plant attached to and drawing nourishment from another living plant.
perennial Living and flowering for more than two years, often flowering each year.
perianth Collective name for petals and sepals where these are not easily distinguishable or collectively for either petals or sepals where only one of these whorls is present.
persistent Remaining attached in situ.
petal Member of the inner series of perianth segments, especially when these are brightly coloured and differ from the outer green sepals of the calyx.
petaloid Having the form, colour or texture of a petal.
phyllaries The small bracts around the outside of a daisy head.
phyllode A flattened leaf stalk which takes the function of a leaf, the true leaf blade being suppressed.
pinnate Of venation, with lateral veins diverging from the mid-rib as do barbs of a feather; of a compound leaf, with leaflets arranged on each side of the mid-rib — also bipinnate, twice pinnate.
plumose Feather-like, with fine spreading pinnate branches.

pod A fruit derived from a single carpel which opens longitudinally along two opposite sides.
pollen The spores containing the male sex cells.
procumbent Prostrate, trailing not rooting.
prostrate Lying closely on the ground, not rooting at the nodes.

raceme A type of unbranched inflorescence having stalked flowers opening in succession, the oldest flowers at the base.
ray floret The outer ligulate or strap florets of a daisy head. These have a perianth as a flap at one side.
recurved Of margins, curved down towards the underside of a leaf; of flower position, curved down.
reticulate In a network, or showing a network of veins.
rhizome A ±underground stem giving rise at the nodes to roots below and shoots above.
rootstock A very short erect underground stem giving rise to roots, buds and shoots.

saprophyte A plant which obtains its food from organic debris in the soil.
scarious Thin, papery or horny, dry, non-green.
sclerophyll A leathery or hard leaf with thick cuticle, hence shrubs and trees with such hard, tough or small stiff leaves.
sepal A unit of the calyx, q.v.
serrate Toothed like a saw blade, the teeth pointed forward.
sessile Without a stalk.
shrub A plant with woody tissue, increasing in diameter annually, branching from the base, usually without a main trunk and smaller than a tree.
sp., spp. Species (singular and plural).
spathulate Shaped like a spoon, broad and rounded at the apex, tapering to the base.
spike An elongated inflorescence of unstalked flowers arranged around an axis, the oldest flowers at the base; the plant may continue to grow beyond the spike as in Callistemon.
spikelet Unit of the inflorescence in grasses; a group of one or more sessile flowers on an axis surrounded by bracts.
spreading Held more or less horizontally away from the axis.
stamen The unit of the male part of the flower, made up of the filament or stalk and the anther which contains the pollen, opening to release it when ripe.
stigma The part of the carpel on which pollen is received.
stipules (adj. *stipulate*) Paired leaf-like or scale-like structures at the base of the leaf stalk.
striate Marked with fine longitudinal lines or ridges.
style The portion of the carpel between the ovary and the receptive surface.
subalpine Areas of relatively high altitude, in Tasmania above ±1300 m.

terete Narrowly cylindrical, without a leaf blade.
throat The opening of a flower tube.
tomentum (adj. *tomentose*) A dense covering of soft matted hairs.
trifoliolate Made up of three leaflets.
tube The cylindrical part of a perianth.

undulate Not lying flat, wavy.
unisexual Of one sex only, unisexual flowers have only functional stamens if male, or only fertile carpels when female.

var. Variety.
venation Arrangement of veins.
vein Strands of conducting tissue often including some strengthening tissue.
verticillate In rings or whorls, as of rings of leaves which arise at the same level on a stem.

whorl A ring of 3 or more similar structures as petals or leaves, arising at the same level on the axis.

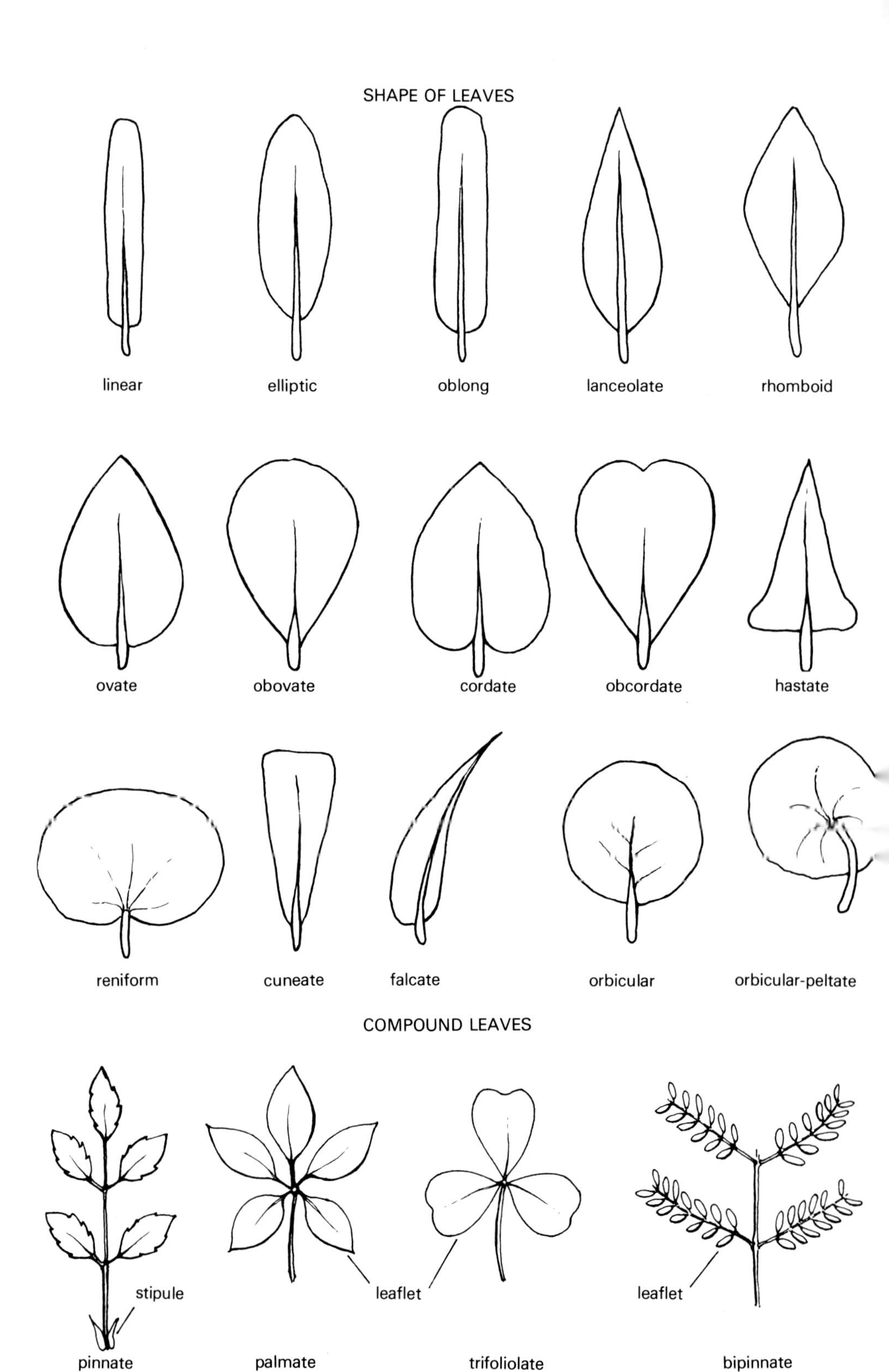
SHAPE OF LEAVES
linear
elliptic
oblong
lanceolate
rhomboid
ovate
obovate
cordate
obcordate
hastate
reniform
cuneate
falcate
orbicular
orbicular-peltate
COMPOUND LEAVES
stipule
leaflet
leaflet
pinnate
palmate
trifoliolate
bipinnate

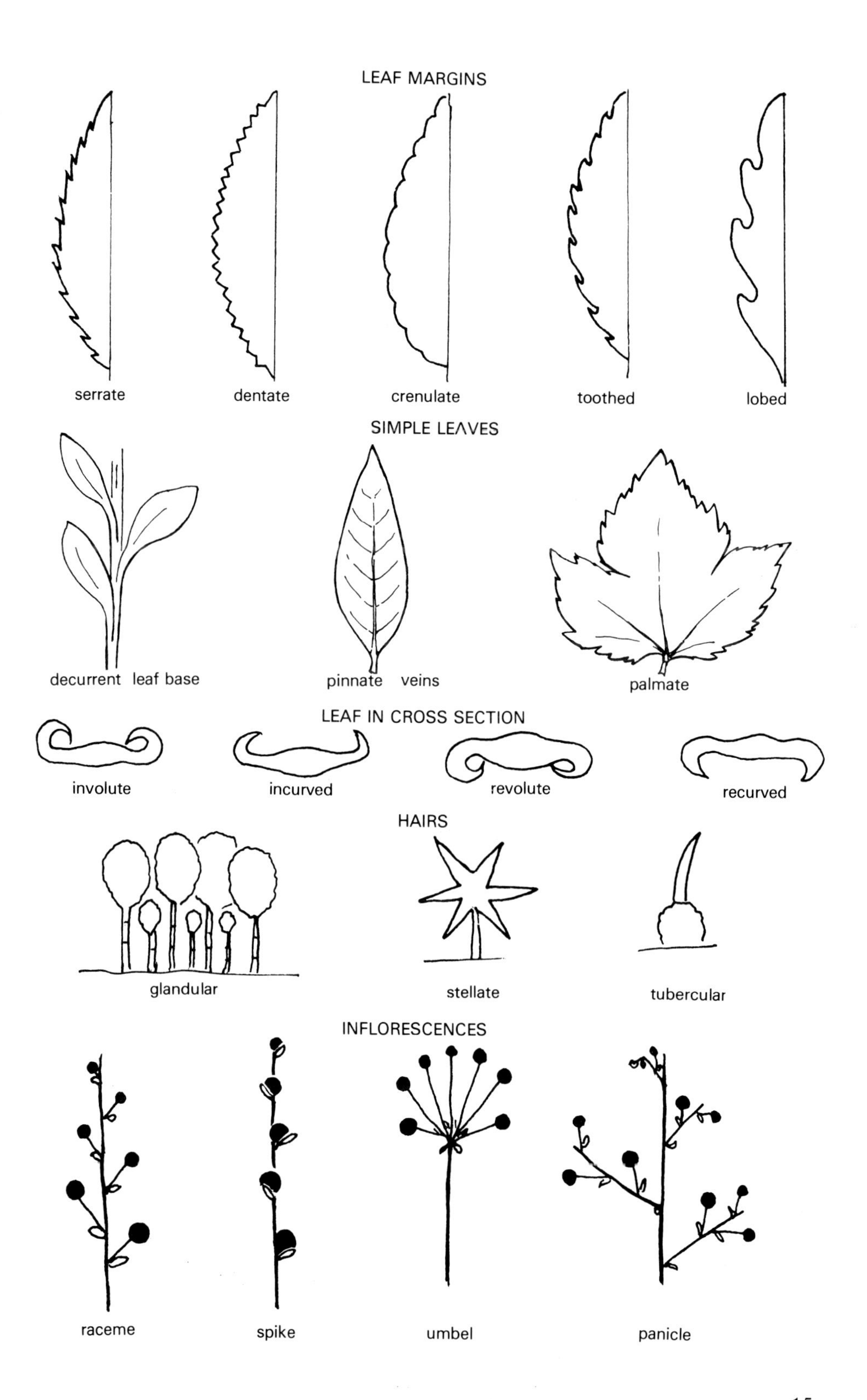
LEAF MARGINS
serrate
dentate
crenulate
toothed
lobed
SIMPLE LEAVES
decurrent leaf base
pinnate veins
palmate
LEAF IN CROSS SECTION
involute
incurved
revolute
recurved
HAIRS
glandular
stellate
tubercular
INFLORESCENCES
raceme
spike
umbel
panicle

THE VEGETATION—GENERAL INFORMATION

Tasmania is a small island of slightly more than 64 000 square kilometres lying in the path of the moisture laden westerly winds. On reaching the west coast these are forced up by the mountain chains, depositing much of their moisture, and in summer having a drying effect on the eastern part of the State. There is a marked rainfall gradient from over 3200 mm in the west near Queenstown to 560 mm on the east coast.

The whole of the western half of Tasmania is consistently wetter than the eastern half. The rainfall ranges from 760 mm in the north-west to 1750 mm in the south-west, while in the eastern half only the north-eastern highlands receive more than 1200 mm — the rest ranging from 800 mm to less than 500 mm in the midlands in the rain shadow of the Western Tiers. The eastern half of Tasmania has higher average summer temperatures and a higher evaporation rate which increases the west-east water availability gradient. This results in marked differences between the vegetation of the west and east coasts. Since good quality forest requires at least 760 mm of rain per annum, this difference permits much larger areas of heavy forest in the west than in the east.

Although so small, Tasmania is very mountainous with over sixty peaks above 915 m and many mountain chains which deflect the prevailing winds causing changes in their humidity and having marked local effect.

The drop in temperature with increasing altitude has a significant effect on vegetation. Places at sea level have six months without frost, but even so low as 305 m frosts may occur at any time of the year; at 610 m agricultural crops cannot be grown in winter and at 915 m almost every day in winter has frost; snow is frequent. While there is no permanent snowline, snow may lie for several months on the higher peaks. The vegetation of these cold areas is restricted to low-growing species which can survive under such severe conditions. They are adapted in various ways which prevent them from freezing and enable them to withstand conditions of low temperatures, cold wet soil and exposure to wind.

The widespread faulting of the dissected and weathered land surface has resulted in sudden changes in rock type and of soils derived from these rocks. There are various soil types including the deep well drained soils of the north west and parts of the north-east which support heavy forest, the fertile soils of the midlands and north which have largely been cleared for agriculture, and the shallow infertile acid soils derived from the siliceous Pre-Cambrian rocks of the south-west which support sedgeland and button grass plains.

At the present time about 2000 species of flowering plants are known to occur in Tasmania, either native to the State or as naturalised introductions. More than 200 native species occur naturally only within the State, that is, are endemic.

The vegetation has two components, the *Antarctic or Southern Oceanic Flora* which shows close affinities with that of South America and New Zealand, and the *Australian* type characterised by wattles (*Acacia* spp.), pea flowers and *Eucalyptus* spp. which Tasmania shares with the rest of south-east Australia.

Many plants regarded as belonging to the Oceanic flora are endemic; many are confined to cold wet situations, their nearest relatives occurring in similar situations in South America, Tierra del Fuego or New Zealand. It is believed that ancestors of these plants existed while Australia, New Zealand and South America were still joined in a southern land mass which included what is now Antarctica.

The Australian flora was shared by the eastern part of mainland Australia and Tasmania during the several times these were joined, but divergence has occurred during times of separation and is still occurring.

Plants which tolerate the same conditions grow together in groups or communities; those easily recognised are salt marsh, montane, and heathland communities. Some occur within several vegetation types wherever there are suitable local conditions — plants belonging to these are widespread. As a result, such plants as the Pink Swamp Heath, *Sprengelia incarnata,* occur from sea level to mountain summit but always in wet conditions in acid waterlogged soil.

The vegetation is constantly changing, and programmes of burning off, land clearance, drainage of swamps, and establishment of certain types of forest are important factors in that change.

Several types of vegetation can be recognised: temperate rainforest or myrtle forest, sclerophyll or eucalypt forest, usually divided into wet sclerophyll and dry sclerophyll, montane vegetation, coastal heath, and sedgeland.

Temperate rainforest or myrtle forest occurs where the rainfall is over 1500 mm spread throughout the year, some falling in summer. It requires constant humidity and grows mainly in the west of the State with smaller areas on the north-eastern highlands. The chief rainforest species are the Myrtle, *Nothofagus cunninghamii,* and the Sassafras, *Atherosperma moschatum.*

Myrtle, with its many small leaves, forms a dense canopy and can shade out other species; fully mature myrtle forest has only a ground cover of ferns and moss. Where the canopy is broken by rivers, fallen trees, roads, landslips or other disturbances, additional rainforest species can grow, including Leatherwood, *Eucryphia lucida,* Celery Top Pine, *Phyllocladus aspleniifolius,* Woolly Tea-tree, *Leptospermum lanigerum,* Pandani, *Richea pandanifolia* and Blackwood, *Acacia melanoxylon.*

Huon Pine, *Lagarostrobos franklinii,* grows along rivers in the west and on swampy flats; other native pines, King Billy and Pencil Pine (*Athrotaxis* spp.) may fringe lakes or form forests on higher slopes. King Billy (*A. selaginoides*) forests still exist on slopes at moderate altitudes; many have been destroyed by fire and their gaunt remains may be seen on many central and western mountains. King Billy is often a constituent of alpine forests with Myrtle or Celery Top Pine.

In places where the water table is high a dense scrub may develop with a mixture of tussocks of Cutting Grass, *Gahnia grandis,* the straggling tough stems of *Bauera rubioides,* sometimes reaching 3 m, the Woolly Tea-tree, *Leptospermum lanigerum,* and the Horizontal scrub, *Anodopetalum biglandulosum* — its slender stems grow upright then fall over and send up upright shoots, which again bend over until the whole is an impenetrable mass of stems.

Where low altitude rainforest is destroyed by fire, eucalypt and blackwood forest may regenerate. Their seedlings require light for growth and grow more rapidly than rainforest species which become understorey trees but may eventually overtake them. Such forest often includes the Manfern, *Dicksonia antarctica,* and merges into wet sclerophyll forest. Myrtle forest is susceptible to wind damage so that ridges and exposed slopes may have eucalypt forest while sheltered gullies and slopes have myrtle forest.

In high rainfall areas if rainforest is burnt repeatedly, wet scrub or button grass sedgeland develops. These fringe areas carry such rainforest understorey shrubs as White Waratah, *Agastachys odorata, Leptospermum lanigerum* and Native Laurel, *Anopterus glandulosus.*

At high altitudes rainforest becomes stunted, giving way to eucalypt woodland with the Snow Gum, *E. coccifera,* passing into montane vegetation.

Montane vegetation occurs on plateaux, mountain slopes and summits. It is characterised by plants which can withstand cold conditions — severe frosts, seasonal snow and strong winds. Although cloudy weather is frequent, there is a high light intensity and occasional very hot days in summer. These conditions cause slow growth and water stress so that the plants are short, often stunted, with small hard leaves and tough-celled wood. The exposure to cold winds cuts young growth so that shrubs are rounded, each shoot protecting the next, a habit which reaches its extreme in the cushion plants. Shrubs often grow in the lee of rocks, protected by them and taking advantage of a rock's ability to absorb heat from the sun.

Five types of community may develop within this vegetation type: dwarf mountain forest, mountain shrubbery, swamp, grassland and mountain fell field.

Dwarf mountain forest may contain conifers such as the prostrate Strawberry Pine, *Microcachrys tetragona,* or the small and erect Cheshunt Pine, *Diselma archeri,* and *Microstrobus niphophilus,* both growing to about 2 m; or perhaps stands of the Deciduous Beech, *Nothofagus gunnii,* a 5 m high tree on sheltered hillsides or a prostrate shrub clinging to the rocky faces of high slopes.

Mountain shrubberies are found in poor rocky soils in exposed situations. They are filled with diverse and interesting plants especially of the daisy, heath and protea families. Shrubs such as *Richea scoparia, R. sprengelioides, R. acerosa* and *Orites revoluta* occur. Members of the daisy family, *Olearia ledifolia, Helichrysum hookeri* and heath family, *Cyathodes* spp. and *Epacris serpyllifolia* are common. Some plants are prostrate and rock hugging such as *Leptospermum rupestre, Baeckea gunniana, Grevillea australis,* others are completely prostrate like *Cryptandra alpina* and *Pentachondra pumila.*

Heathlands and grasslands with *Poa* and *Carpha* occur on ridges and better drained slopes, with shrubs like the Lemon-scented Boronia, *Boronia citriodora* and berry bushes, *Cyathodes* spp. Herbs such as buttercups, native dandelion, *Microseris, Euphrasia* sp. and daisies of *Senecio* spp. occur.

Swamp areas may be dominated by sedges such as *Restio,* with sphagnum moss throughout, and prickly heath plants around the edges.

In the flatter areas bordering swamps or near mountain summits where snow lies for several months communities of very short plants develop, including the cushion plants of which Tasmania has five species. The sage green *Pterygopappus lawrencii* with rather square stem tips is easily identified, but the other four form dark cushions very similar in appearance when not in flower. They are shrubby plants, prostrate, with very many short parallel erect shoots, tightly packed and laced together by roots to make a rounded mound so firm that it does not dent when walked on. Such plants increase in diameter and may coalesce forming large mounds. They grow across small watercourses, impeding drainage and slowing run off, helping to prevent erosion. Seeds of other plants may become established between the leaves of the cushion and grow with their roots inside it — such plants include *Nertera, Pernettya, Drosera arcturi, Sprengelia incarnata* and *Plantago gunnii.*

Some plants such as the orchid *Prasophyllum suttonii* and herbaceous *Senecio* spp. escape the cold by spending the winter in an inactive form, as tubers or rosettes closely pressed to the ground, then grow rapidly and seed, dying back in the autumn.

Wet sclerophyll or eucalypt forest occurs on deep fertile soils where the rainfall is between 1000 mm and 1500 mm per year. The chief trees are the valuable hardwood species *Eucalyptus delegatensis, E. regnans, E. obliqua, E. sieberi.* The species vary with locality; *E. ovata* is important near Smithton where it reaches a large size; *E. globulus* is important in the south.

The open canopy allows the development of a shrubby understorey, and in sheltered gullies where humidity is high, rainforest species will grow — Myrtle, Sassafras, Silver Wattle, *Acacia dealbata,* with the Tree Fern, *Dicksonia,* Musk, *Olearia argophylla, Bedfordia* and Dogwood, *Pomaderris apetala.* Understorey trees such as Christmas Bush, *Prostanthera lasianthos, Pittosporum bicolor,* Stinkwood, *Zieria arborescens,* and in some high altitude forests Waratah, *Telopea truncata,* and *Leptospermum lanigerum* occur.

Clearings are filled with shrubby species like Varnished Wattle, *Acacia verniciflua, Oxylobium ellipticum,* Prickly Beauty, *Pultenaea juniperina.* Cold wet flats, after being burnt, may support grassland or sedgeland and shrubs like the Mountain

Berry, *Cyathodes parvifolia* and the Mountain Pepper, *Tasmannia lanceolata*.

With decreasing rainfall wet sclerophyll passes into a more open forest, the dry sclerophyll forest, the most widespread eucalypt being the endemic Black Peppermint, *Eucalyptus amygdalina,* with *E. viminalis,* the White Gum. Other species are rather local, changing with soil type; the Cabbage Gum, *E. pauciflora,* is widespread especially in the midlands, central south and east, while the silver-leaved *E. tenuiramis* is common on the drier mudstone soils of the south-east.

Wattles, *Banksia marginata,* Bull Oak, *Allocasuarina littoralis,* Native Cherries, *Exocarpos* spp. and Prickly Box occur as understorey trees; there is often a lower shrub layer of pea flowers, heaths, *Leucopogon* and *Epacris* and *Tetratheca* spp. Open grassy slopes occur with herbaceous plants including orchids, and rocky hillsides sometimes support almost pure stands of the She-Oak, *Allocasuarina verticillata.* Wet areas carry thicker vegetation and wet flats have sedges of the *Lepidosperma* spp. and Cutting Grass *Gahnia grandis.*

Sedgeland occurs where heath, scrub, or forest has been repeatedly burnt, or on poor peaty, acid soils where the water table is high. Large areas of sedgeland with Button Grass, *Gymnoschoenus sphaerocephalus* occur in the south-west on infertile poorly drained soil derived from Pre-Cambrian rocks, or on the waterlogged flats in valley bottoms. Sedgelands occur in other places in wet waterlogged soils as on mountain moors and coastal heaths. Sedgeland plants have special adaptations to cope with such poorly aerated soils.

Coastal heath is most extensive in the far north-west, north-east and east and near Strahan, and on islands of Bass Strait. It occurs on sandy soils, often developed from wind blown sand. Such soil is low in minerals needed for plant nutrition. The characteristic heath vegetation consists of shrubs less than 2 m high with hard or leathery leaves.

The trees are small, stunted by strong wind; they include *Eucalyptus amygdalina* in the east, *E. nitida,* in the west, *Melaleuca* spp., *Banksia marginata,* Sunshine Wattle, *Acacia terminalis* (*A. botrycephala*) and *Allocasuarina monilifera* or in the west *A. zephyrea.*

The winter in such areas is less severe than farther inland and the soil warms quickly in spring giving a flush of flowering in September to November. Many small herbs and bulbous rooted plants occur. These are able to flower early under the mild spring conditions and die down in the heat of summer.

Shrubs common in the sandy heaths are *Pimelea linifolia, P. flava,* plants of the heath family, Epacridaceae, including *Leucopogon virgatus, L. collinus, Epacris impressa,* pea flowers *Aotus ericoides, Pultenaea stricta, Bossiaea cinerea,* Myrtaceae such as Bottlebrush, *Melaleuca* spp., Tea-tree, *Leptospermum scoparium,* the grass trees *Xanthorrhoea australis* and *X.* sp., *Hakea* spp. varying with locality, several species of *Hibbertia, Tetratheca,* and *Comesperma.*

In winter the soil may become waterlogged, and patches of Button Grass, with species of Bladderwort, *Utricularia,* Sundew, *Drosera* and *Xyris,* occur. The drainage is often poor due to formation of hard pans or it is impeded by sand dunes; fresh or brackish lagoons and reedy swamps are common with a wide variety of water plants.

Heathland is maintained by burning; if no fire occurs the shrubs become taller forming a scrub forest or eventually an open woodland. Heath species show various adaptations which enable them to survive fire; very many send up new shoots from bulbs, rhizomes or woody knobbed rootstocks buried just beneath the damp soil; some like *Banksia* and *Hakea* produce woody fruits which protect the seed during fires and need heat to dry and open them, others such as the pea flowers and wattles produce seeds which need to be heated by fire before they can germinate. Coastal heaths adjoin sand dunes, coastal cliffs and salt marshes, and plants from these areas have been included in the section dealing with coastal plants.

Mary Cameron

MONTANE

Plants of mountain areas are subject to harsh conditions for several months. No definite altitude can be given as the lower limit for montane vegetation, for its development often depends on local conditions of temperature and exposure.

Nothofagus gunnii *Deciduous Beech, Tanglefoot*

This small tree is the only native tree whose leaves change colour and fall completely each autumn. Many people travel to Cradle Mountain and other central and western mountains at the end of April to enjoy this sight. The beech is a small tree 1.5-5 m depending on severity of conditions. Its leaves are shortly stalked, ovate or almost orbicular, folded concertina fashion parallel with the pinnate veins, margins serrate. Flowers small, male flowers with pendent stamens, female with protruding styles, wind pollinated. Fruit small nuts in woody fruits nearly 1 cm long with 4 valves made up of overlapping scales. Flowering December. Western and southern mountains, 1000-1400 m. Tas, endemic.

Fagaceae

Baeckea gunniana

An aromatic spicy scented shrub with very narrow crowded leaves dotted with oil glands. It is found on mountain plateaux where it is often prostrate and rock hugging. Leaves narrow-oblong, 2-6 mm long, thick, almost filiform, dark green. Flowers small, white, 4 mm across, 5-petalled, very numerous towards the ends of branches, stamens 5 or more around the edge of the nectar disc between the petals. Fruit capsular. Flowering January-February. Common on mountains above 800 m. Tas, Vic, NSW.

Myrtaceae

Podocarpus lawrencii *Plum Pine*

A small tree at moderate altitudes and in sheltered positions but reduced to a low ascending shrub between the boulders near the summits of mountains. The 'leaves' are flattened stalks, 6-12 mm long, dark green, linear-oblong, flat, hard and leathery. Male cones are small pinkish-red, rather long, one or two together on very short stalks in terminal axils. Female cone of a single stalked fleshy scale carrying a blackish green seed. When mature the seed scale and its stalk become red and succulent. On stabilised talus at altitudes between 1000 m and 1500 m and at lower altitudes along the banks of some rivers such as the Mersey. Tas, Vic, NSW.

Phyllocladaceae

Eucalyptus coccifera *Tasmanian Snow Gum*

Usually a small tree growing in rocky doleritic sites exposed to wind and snow, this Eucalypt has smooth white and grey bark and yellowish or reddish twigs. Juvenile leaves opposite, ovate and grey-green, on papillose twigs. Adult leaves green or greyish, about 8 cm long, elliptical-lanceolate, with a small hooked tip (often missing). Flower buds long, ridged, in umbels of 5-7 (3-flowered on Mt Wellington). Fruit large, 8-14 mm across, glaucous, smooth or ridged, hemi-spherical or deeply cylindrical bowl-shaped, disc at top flat, often dark coloured. *E. coccifera* may grow to 40 m under very favourable conditions. Very resistant to cold. Flowering usually January, occasionally other months. Subalpine on exposed mountain tops and plateaux at altitudes between 800 m and 1350 m, widespread except in north-east. Tas, endemic.

Myrtaceae

1 *Nothofagus gunnii*
Deciduous Beech at Crater Lake

2 *Nothofagus gunnii*
Deciduous Beech, Tanglefoot

3 *Baeckea gunniana* Baeckea

4 *Podocarpus lawrencii* (fruits) Plum Pine

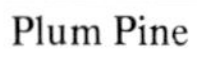

5 *Eucalyptus coccifera*
Tasmanian Snow Gum

6 *Eucalyptus coccifera* (juvenile leaves)
Tasmanian Snow Gum

Olearia tasmanica
Formerly known as *O. alpina*. An erect shrub, much branched in the upper part. Leaves elliptical, blunt, 1-2 cm long, dark green and glossy above, fawn-brown beneath. Flower heads large long-stalked, solitary in axils near the ends of branches. 5-6 white ray florets, disc florets yellow. Pappus bristles, long straight, light brown. Flowering January. Alpine, in shrubberies and wet scrub, to sea level in south-west. Tas, endemic.
Asteraceae

Tetracarpaea tasmanica
A small shrub with many branches to 1 m. Leaves dark green, about 2 cm long, elliptical, thick, shining, with toothed margins, upper surface with numerous conspicuously indented veins. Flowers with 4 white spreading sepals and 4 white erect petals. Small 6 mm flowers are crowded in terminal racemes 4 cm long, often several flowering shoots parallel and close together. Fruit long narrow angular capsules, usually 4 together on the fruiting raceme. Flowering November-December. Locally frequent in montane shrubberies. The only species in this genus and endemic to Tas.
Escalloniaceae

Orites revoluta
A dense bushy shrub of mountain tops and plateaux with a dense covering of short brown felted hairs on all the younger parts. Leaves elliptical, dark green, tough, leathery, to 2 cm, margins strongly revolute obscuring most of the underside which is covered with short dense brown hairs. Flowers creamy yellow, paired in short terminal spikes to 4 cm long, the buds cylindrical with expanded ends. As flowers open perianth splits into four narrow segments each bearing a shortly stalked stamen near the base of the expanded part. Fruit a short blunt follicle, about 1 cm long, covered with dense brown hair. Fruit winged. Flowering January. All four species of *Orites* in Tasmania are endemic and part of the alpine flora. There are five other species of which four are found in Eastern Australia and one in Chile.
Proteaceae

Olearia ledifolia
A densely branched small shrub with tough, leathery, strongly revolute leaves 12-14 mm long. Young branches, flower buds, stalks and undersides of leaves covered with a felt of short brown hairs. Leaves are carried for 1-2 years, then fall leaving the lower branches bare but with rough raised scars which remain till shed with the bark. Flower heads are large white daisies more than 1 cm across on stalks 1 cm long, solitary in axils near ends of branches. Outside bracts of each flower head are covered with rusty brown hairs, the ray florets long and white, the disc florets yellow. Pappus, straight rust-coloured hairs. Flowering January-February. Mountain shrubberies and among boulders on mountain plateaux. Tas, endemic.
Asteraceae

Boronia citriodora *Lemon-scented Boronia*
A small shrub less than 1 m high, distinguished from other Boronias by the strong lemon scent of its crushed foliage. Leaves erect, opposite, pinnate with 3-7 linear leaflets 5-15 mm long. Flowers with 4 spreading whitish pink petals, the under surface often darker pink or red especially in the bud stage. Stamens 8. Fruit separating into 4-seeded parts. Flowering December-March. Common in subalpine heaths and at sea level in the south-west. Tas, endemic.
Rutaceae

Microcachrys tetragona *Creeping Pine, Strawberry Pine*
This prostrate creeping pine appears to have square stems because the tiny thick opposite leaves overlap closely, each pair fitting into the notches between the previous pair as if in a square plait. Male and female cones are on different plants, the male very small at the end of the branchlet, little wider than the foliage. The female cone of about 24 scales is ovoid, a little less than 1 cm long, red and fleshy when ripe, the seeds buried between the red scales, one seed per scale. Ripe cones are found in January-February. On central, western and south-western mountains between 1350 m and 1500 m, on ridges and in high moorland. Tas, endemic.
Phyllocladaceae

7 *Olearia tasmanica*

8 *Tetracarpaea tasmanica*

9 *Orites revoluta*

10 *Olearia ledifolia*

11 *Boronia citriodora*
Lemon-scented Boronia

12 *Microcachrys tetragona* (fruits)
Creeping Pine, Strawberry Pine

Richea scoparia
A shrub to 3 m where sheltered, but often a dense rounded wind-pruned bush 40-100 cm. Inflorescences colourful, formed by terminal spikes to 12 cm long of flowers with caps of joined petals in shades of orange, yellow, red, pink or white. The petals are deciduous, exposing 5 mm long stamens and a short style. Leaves are hard, sharp-pointed, linear-lanceolate with a sheathing base, crowded, remaining on the plant for several years even when dead. One of the most conspicuous plants in mountain shrubberies and on plateaux. Flowering January. Widespread and abundant on mountains. Tas, endemic.
Epacridaceae

Cyathodes straminea
A much branched spreading shrub, leaves short up to 1.5 cm clustered in apparent whorls towards the ends of the branches, the twigs bare between successive groups of whorls. Leaves elliptical, blunt, lower surface paler and striate. Flowers creamy white, long-tubular, of 5 joined petals, stamens near mouth of tube. Fruit a flattened red drupe. Distinguished from the larger *C. glauca,* Cheeseberry, by its shorter more bluntly pointed leaves. Flowering December-January. Common on summits of most mountains. Tas, endemic.
Epacridaceae

Leptospermum rupestre
Usually prostrate and rock hugging but sometimes an erect shrub to 2 m high. Shortly stalked leaves are ovate, 5-8 mm long, concave, smooth and shining, slightly aromatic and densely crowded. Flowers are white, 12-15 mm in diameter, in leaf axils and terminal on short lateral branches. Each flower has 5 sepals, 5 longer orbicular petals, numerous stamens in a ring around the central disc. Fruit, a capsule about 5 mm diameter, top convex, woody when mature, opening by 5 slits. Flowering January. Abundant on mountains, 900 to 1500 m. Tas, endemic.
Myrtaceae

Milligania densiflora
Tufted plant with wide grass-like leaves 15-30 cm long, 2-2.5 cm wide, clustered at the base of the flowering stalk. Flower stalk covered with felted hairs; underside of leaf base, floral bract and perianth densely hairy. Flower stalk 15-30 cm high, bearing one or two bracts, inflorescence branched. Flowers cream, numerous, tubular with 6 blunt lobes about 1 cm long, 6 stamens and 3-lobed ovary. Flowering November-December. A plant of mountain hillsides and rocky banks. Tas, endemic.
Liliaceae

Cyathodes petiolaris
A low bushy shrub 15-30 cm high, branches spreading. Leaves dark green, stalked, with shining upper surface, lower surface conspicuously striate, visible because the leaves are semi-erect. Flowers small, solitary or 2-3 in leaf axils, white or pinkish. Fruit a small red drupe, flat or hollow-topped. Flowering December-January. Abundant and widespread on mountain plateaux above 1100 m. Tas, endemic.
Epacridaceae

13 *Richea scoparia* with Pencil Pines

14 *Richea scoparia*

15 *Cyathodes straminea*

16 *Leptospermum rupestre*

17 *Milligania densiflora*

18 *Cyathodes petiolaris*

Phebalium montanum
A low mountain shrub to 60 cm high or almost prostrate. Leaves to 12 mm long, crowded, dark green, linear, fleshy, blunt, marked with glandular dots. Stems dark with rough leaf scars. Flowers solitary, white or pinkish, shorter than the leaves, up to 1 cm long, sepals 5, petals 5, cream tipped with crimson; stamens usually longer than petals, anthers crimson before pollen is shed. Flowering November-January. Found in alpine and subalpine areas in north. Tas, endemic.
Rutaceae

Epacris serpyllifolia
A white-flowered, much branched heath to 80 cm high in sheltered places but prostrate forming a dense mat in exposed areas. Leaves shining, dark green or reddish, ovate, densely crowded, pointed but not sharp. Leaf stalks thick, leaves 2-5 mm long, keeled beneath or with three longitudinal furrows. Flowers large, solitary in leaf axils near the ends of branches forming dense heads. Bracts and sepals papery, straw coloured or grey with white margins. Corolla tube white about the same length as calyx, the lobes short. Stamens at mouth of tube. Flowering spring-early summer. Montane moorlands to mountain summits. Tas, Vic.
Epacridaceae

Pimelea sericea
A small shrub of mountain moorlands with many erect branches and all parts except the upper surfaces of the leaves silky-hairy. Leaves shortly stalked, broad-elliptical, 7-12 mm long, usually opposite and crowded at the ends of branches. White to pink tubular flowers in terminal heads surrounded by bracts similar to the leaves. Separate male and female flowers in the one head with female flowers slightly shorter than the 8-10 mm long males. Bases of the individual florets surrounded by short hairs. The small dry fruit has a tuft of long hairs at the top. Flowering October-January. Tas, endemic.
Thymelaeaceae

Richea sprengelioides
A stiff woody shrub to 1 m, young branches rough with encircling scars of fallen leaves, older branches with grey flaking bark, becoming smooth grey. Leaves lanceolate, crowded, with wide sheathing bases wrapped round the stem, blades hard, spreading, parallel veined, rather wide, tapering quickly to a hard point, more or less in 4 rows. Flowers pale yellow in small terminal heads hardly wider than the leaves. Each flower with its bracts solitary in a leaf axil, perianth cap yellow, soon shed. Stamens long, thick, anthers oblong. Fruit capsular. The foliage resembles that of *Sprengelia incarnata* but the plant is far more substantial, the leaves are wider with a much wider angle at the point. Flowering November-December. Widespread on mountains. Tas, endemic.
Epacridaceae

Rubus gunnianus
A small alpine perennial herb with rosettes of stalked usually trifoliolate leaves and white flowers, producing new plants from underground runners. Leaves dark green, shining, leaflets lobed and margins serrate; flowers shortly stalked with 5 sepals, 5 white narrow tongue-shaped petals, stamens numerous. Fruits 1 cm diameter, 4-5 red juicy segments clustered as in raspberry. Edible. Flowering November-December. Common in open mountain areas 1000-1750 m. Tas, endemic.
Rosaceae

Anemone crassifolia
Small white-flowered herbaceous plant growing from a thickened rootstock. Leaves dark green, often purplish beneath, reniform but palmately divided into 3-5 coarsely toothed lobes. All leaves long stalked, arising from base of the stem. Flowering stalk erect 8-30 cm long, with 3 small lobed bracts towards the top. Flower large 2-3 cm across, solitary, terminal, consisting of 6-8 white petaloid sepals, numerous stamens in many rows and numerous one seeded ovaries. Fruit, dry with a hooked or coiled style. Flowering December-January. Local on mountains in the west at about 1350 m. Tas, endemic.
Ranunculaceae

19 *Phebalium montanum*

20 *Epacris serpyllifolia*

21 *Pimelea sericea*

22 *Richea sprengelioides*

23 *Rubus gunnianus*

25 *Anemone crassifolia*

24 *Rubus gunnianus* (fruits)

Drosera arcturi *Alpine Sundew*
An alpine insectivorous plant growing on wet ground in bogs and in cushion plants. Roots fibrous, without tubers. Leaves erect, narrow, strap-shaped, about 10 cm long 6 mm wide, few together in a basal rosette, light green or bronzed. Large glandular hairs on upper half of leaf and around margin. Flowers solitary, large, 2 cm across, white or cream on erect stalks slightly longer than leaves. Petals about 10 mm long, obovate; ovary large, stigmas conspicuous. Fruit ovoid capsule often covered by blackish remains of the sepals. Flowering December. Tas, Vic, NSW; New Zealand.
Droseraceae

Coprosma moorei
A hairless alpine plant, with thin prostrate creeping stems and small erect shoots; stems rooting, forming small mats. Leaves opposite, ovate or lanceolate, pointed, thick, shining, about 4 mm long. Flowers bisexual (male and female in same flower), joined sepals, petals forming a short tube, 4-5 stamens and 2 stigmas protruding from the tube. Fruit a blue spherical 5 mm diameter drupe, 2-seeded. Flowering November. Montane and alpine peaty heaths. Tas, Vic.
Rubiaceae

Donatia novae-zelandiae
Probably the commonest cushion plant in Tasmania forming hard dark green cushions of tightly packed branching leafy shoots. Cushions are hard enough to take a man's weight without fracturing and are held together by interlacing roots from buried leaf axils. In spite of this, seedlings of many alpine plants manage to grow in the tops of cushion plants. Leaves small, pointed, closely overlapping on each individual stem, with tufts of white hairs at the base of the leaf. Flowers solitary with small calyx and 5 white fleshy petals 5 mm wide, 2 stamens joined by their bases to disc near style. Fruit not opening. Flowering December-February. Exposed mountain summits and plateaux, in cold wet areas. Tas; New Zealand.
Donatiaceae

Ewartia catipes
A very small alpine daisy, perennial with many prostrate rooting branches to 15 cm long forming silvery mats. Leaves obovate, with wide stalks, blade folded about the mid-rib, silvery-white, with short dense silky hairs. Flowering heads in small branched terminal clusters; the whole cluster surrounded by the subtending leaves and at first glance looking like a minute posy. Each small head of the cluster has white outer bracts and up to 10 tiny crimson or white disc florets. Fruits with pappus. Flowering December. Exposed places on north-eastern, central and western mountains. Tas, endemic.
Asteraceae

Ewartia meredithae
A dwarf herbaceous daisy, forming cushion-like mats of erect branched leafy stems in alpine situations. Plants 2-6 cm high, leaves about 5 mm long spathulate (spoon shaped), erect, more or less parallel with stem, grey-yellow, or rusty with short soft hairs. Flower heads terminal, unstalked and level with leaves, or erect on longer stalks. Head a small white paper daisy with brown outside bracts, conspicuous narrow white spreading bracts and small whitish tube flowers in the protruding eye of the daisy. Fruit with pappus. Flowering December. Wet exposed places on mountains, but very local. Tas, endemic.
Asteraceae

Gaultheria depressa
Small woody shrub with fairly stout stems, much branched, older branches grey, but young stems brown clothed with stiff appressed brown bristly hairs. Leaves broadly elliptical to orbicular, 6-8 mm long, surface reticulate, shortly stalked, margins toothed, each tooth tipped with a short stout bristle. Flowers axillary near the ends of branches, sepals 2 mm, petals about twice as long. Ovary 5-lobed, style short straight. Fruit a capsule surrounded by the enlarged red fleshy calyx. Flowering December-January. Occasional on mountains. Tas; New Zealand.
Ericaceae

26 *Drosera arcturi* Alpine Sundew

27 *Coprosma moorei* (fruits)

28 *Donatia novae-zelandiae*

29 *Ewartia catipes*

30 *Ewartia meredithae*

31 *Gaultheria depressa* (fruits)

Montia australasica
A hairless prostrate creeping or straggling herb of wet places. Leaves alternate, linear to 3 cm long, flat, slightly fleshy, veins not visible, with papery sheathing bases. Flowers stalked, 2 sepals, 5 free white petals, 5 stamens opposite the petals, and attached to them at the base, 3 joined carpels with spreading styles. Fruit a capsule. Flowering spring and summer. Forming short dense mats in wet flats or long stemmed and straggling through taller vegetation in ditches and soaks. Sea level to montane. Tas, temperate mainland; New Zealand.
Portulacaceae

Carpha alpina
A small grass-like sedge forming tufts and tussocks in montane and alpine regions. Leaves narrow-linear, with shining straw-like brown sheathing bases and channelled blade 5-15 cm long, 2 mm wide, shorter than the stems. The erect stems bear 1 or 2 clusters of spikelets each cluster subtended by a long narrow bract; 3 or 4 pale flattened spikelets together in stalked heads, 2 or 3 heads in each cluster. Each spikelet has 2-3 chaffy bracts surrounding a single flower with 6 long scales each plumose with long silky hairs, 3 stamens and an angular ovary with a long straight style. The feathery heads make this plant easy to identify. It is found on most mountains. Tas, Vic (rare); New Zealand and New Guinea.
Cyperaceac

Pernettya tasmanica
A small woody shrub with creeping rooting stems and many spreading or ascending wiry branches. Leaves alternate, without hairs, narrow elliptical, 4-6 mm long, 2-3 mm wide, flat or concave, margins with 1-3 pairs of minute teeth. Flowers white, cup-shaped, 4 mm across, sepals 3 m long, persistent; petals joined at base, lobes narrow, soon shed. Stamens 4 mm long with short anthers; style short, straight. Fruit very large for size of plant, nearly 1 cm across, red or occasionally yellow, slightly 5-lobed, globose but depressed with central remains of style, sitting on fleshy calyx the same colour, like cup on saucer. Flowering December-January. Local in mountain moorlands and on mountain tops. Tas, endemic.
Ericaceae

Euphrasia collina ssp **diemenica**
This *Euphrasia* with its large white 2-lipped flowers is common at Mt. Barrow and some other mountains. The flowering stems die after flowering leaving a few short vegetative stems with smaller leaves attached to the persistent rootstock. Flowers in this form are usually pure white with large yellow blotches about the lower lip; leaves variable in shape with 3-7 lobes. Flowering late spring to summer. Mountain plateaux, in heathland and grassy areas. See also plate 47. Endemic.
Scrophulariaceae

Cryptandra alpina
Small is beautiful! Prostrate woody plant with fine wiry stems and minute leaves 2-4 mm long, creeping over alpine meadows and at bases of rocks, rooting at intervals along stem. Leaves very fine, only 0.5 mm wide, stalked, stipulate, finely pointed. Flowers 1 or 2 together at ends of branches, surrounded by bright brown bracts. Flowers shortly tubular, with 5 short white spreading lobes, 5 minute white erect petals, forming hoods over the stamens; ovary enclosed. Fruit a capsule. Flowering December-February. Locally abundant on Central Plateau and Western Tiers above 1000 m. Tas, endemic.
Rhamnaceae

Dracophyllum minimum
A member of the heath family with very numerous, parallel short branches massed together to form a large hard dark green cushion-like mound. When not flowering it is difficult to distinguish from *Abrotanella forsteroides*. The leaves are 3-6 mm long, dark green, narrow, sharply pointed, stem-clasping, with minutely rough edges. White flowers 6-11 mm in diameter resembling *Epacris* flowers with frilly petals, are terminal on the shoots, their bases embedded in the top of the cushion. Fruit ± spherical, crimson, fleshy at first, then dry, splitting. Flowering spring-summer. At high altitudes on mountains of Central Plateau and west coast. Tas. endemic.
Epacridaceae

32 *Montia australasica*

33 *Carpha alpina*

34 *Pernettya tasmanica* (fruits)

35 *Euphrasia collina* ssp *diemenica*

36 *Cryptandra alpina*

37 *Dracophyllum minimum*

Celmisia saxifraga *Small Snow Daisy*
A herbaceous perennial daisy covered with white silky hairs. Leaves short 1.5-3 cm long in a radical rosette, lateral branches forming clumps of rosettes in older plants. Leaves narrow-elliptical, yellowish or grey-green above, yellowish-white below, ridged longitudinally. Flower head white or purplish-white, with yellow centre, solitary on white stalk to 15 cm long, bearing several small narrow bracts. Phyllaries green, herbaceous. Whole plant sometimes tinged with purple. Flowering January. Slopes of mountains where snow lies. Tas, endemic.
A second species *C. asteliifolia*, a larger plant with longer leaves, is very similar but is more abundant and reaches down to lower altitudes in montane grasslands. Flowering January. *C. asteliifolia* occurs in other States.
Asteraceae

Erigeron pappochroma
A small herbaceous daisy, all parts rather glandular, with a rosette of stalked spathulate leaves 1-7 cm long. One variety has rather rough leaves with sparse short bristly hairs, another is softly hairy. Flower stalks erect with 1 or 2 linear bracts. Bracts around flower head herbaceous, ray florets many, deep or pale mauve, narrow and often infolded, disc florets numerous. Several distinct forms have been recognised and given varietal status. Flowering summer. Subalpine in grassy places. Tas, Vic, NSW.
Asteraceae

Astelia alpina *Pineapple Grass*
The tufted rosettes of stiff leaves give this member of the lily family the name Pineapple Grass. It forms large mats of close growing plants in bogs and around runnels on mountain moors, spreading by rhizomes. Leaves linear, channelled, green above and silver beneath, about 20 cm long. Flowers greenish in terminal spikes in the centre of the leaves which continue to grow up around them so that the fruit is hidden. Fruit, mucilaginous bright red ovoid berries 12 mm long hidden in the heart of the tuft. Berries edible. Seeds 5, black shining. Flowering May. Wet places, bogs, around water courses on mountains. Tas, Vic, NSW.
Liliaceae

Dracophyllum milliganii
Woody plant branching only at base, with erect stems ending in crowded spiral tufts of leaves, or leaves and an erect flowering head. Height 15 cm to 4 m, the smaller plants being in more open positions so more often seen. Leaves long, 15-60 cm, hard, stiff, tapering from broad sheathing base to a long tip, margins rough. Flower head large to 30 cm long, the axis with leaf-like bracts diminishing in size towards the tip, each bract subtending one or more long branched spikes of wide-tubed fleshy heath-like flowers. Both petaloid sepals and petals white or pink; stem and bases of bracts deep red. Fruit a capsule. Flowering December-January. Wet areas of south-west Tasmania in sheltered gullies and on exposed ridges and slopes. Tas, endemic.
Epacridaceae

Isophysis tasmanica
An iris-like plant with fans of short linear leaves about 15 cm long, arising from a branched woody rhizome. Leaves parallel veined, linear and stiff, overlapping in two rows, green with conspicuous brown margins. The flower, borne on a short stalk 15-30 cm long, is solitary arising between two long papery bracts. Each flower is surprisingly large, 8 cm across, either very dark purple-red or yellow, with 6 narrow-lanceolate spreading 'petals' joined only at the base, 3 stamens and an ovary half embedded in flower tube. Fruit capsular, seeds numerous. Flowering December-January. Mountains near the west coast and at sea level, in wet moor-lands in west and extreme south-west. Tas, endemic.
Iridaceae

Helichrysum pumilum
A small perennial mountain everlasting, leaves linear about 4 cm long in a rosette at base of the stem. Flower stalks to 10 cm high, whitish with a few linear bracts; flower head 1.5-2 cm across, solitary, terminal, outer bracts purplish red or brown, inner ones white with wide bases. Pappus bristles minutely rough with thickened tips. Flowering December-January. Mountain herbfields reaching down to coast in south-west and west. Tas, endemic.
Asteraceae

38 *Celmisia saxifraga* Small Snow Daisy

39 *Erigeron pappochroma*

40 *Astelia alpina* Pineapple Grass

41 *Dracophyllum milliganii*

42 *Isophysis tasmanica*

43 *Helichrysum pumilum*

Gentianella diemensis *Gentian*
An alpine herb about 30 cm high with a rosette of wide-stalked ovate or lanceolate leaves and sessile opposite stem leaves. Stems often purple-green to almost black. Flowers conspicuous, 5-petalled, cup shaped, pure white or more usually striped with purple. Flowering spring-summer. Common in alpine and subalpine grasslands but also local on hills at lower levels. Tas, Vic, NSW, SA.
Gentianaceae

Helipterum albicans
A white-flowered perennial everlasting daisy with a stout woody rootstock. Stems branching, leafy at the base, young stems white with woolly hairs. Leaves alternate, narrow linear, almost flat or with recurved margins 4-12 cm long, both surfaces softly felted, white. Flower heads solitary, terminal on long stems with a few leafy bracts. Flowers white or tipped with reddish purple 2.5-3.5 cm wide. Flowering December-January. Locally frequent in midlands and north-west, sea level to montane grasslands.
On the mainland a yellow-flowered form is known as *H. albicans.* The white-flowered form sometimes known as *H. albicans* var. *incanum,* is confined to Tasmania. Tas, endemic.
Asteraceae

Herpolirion novae-zelandiae *Sky Lily*
A small rhizomatous perennial blue-flowered lily of alpine herbfields, creeping and forming grass-like mats in swampy areas. Leaves linear, stiff, grey-green, 3-4 cm long, 2-4 mm wide. The shortly stalked flowers nestling among the leaves are blue or bluish-white, large, 2 cm across, with 6 stamens and 3-celled ovary. Capsule spherical, with few smooth black seeds. Flowering December-January. Common in herbfields in central and western mountains. Solitary member of the genus Herpolirion. Tas, Vic, NSW; New Zealand.
Liliaceae

Euphrasia collina ssp **diemenica** *Eyebright*
Perennial with a woody base, some vegetative stems often prostrate and rooting, with crowded opposite leaves forming 4 rows. Flowering stems erect, stout, often purplish in colour, dying after flowering. Flowers large, lilac or white in short terminal racemes, usually all open flowers and buds at about the same level, intervals between them increasing as seed sets. Bracts and calyces usually with short glandular hairs. Tube of flower 1½ times length of calyx; flower wide, anthers reddish-brown. Flowering summer. Locally abundant on mountains descending to the coast only in south and west. Tas, endemic.
Scrophulariaceae

Gaultheria hispida *Snow Berry*
Erect spreading shrub 1-2 m high, branches and mid-ribs of leaves covered with coarse brown hairs. Leaves shortly stalked, elliptical to lanceolate, pointed, margins serrated. Small urn-shaped white flowers in terminal clusters; stamens 10. Calyx enlarges to form a fleshy white or pinkish-white globe around the dry capsular fruit. Flowering November-December. Frequent in rainforest areas and wet eucalypt forests, sea level in west to montane. Tas, endemic.
Ericaceae

Bellendena montana *Mountain Rocket*
Small subalpine shrub 10-60 cm high with erect or low spreading branches. Leaves stalked, obovate or cuneate, 2.5 cm long, entire or with 3 blunt lobes bright green or blue-green. Cream flowers in pyramidal terminal racemes 4-6 cm long, on 4-6 cm stalks. Individual flowers have 4 segments 3-4 mm long, at first spreading, then curling back. Fruit brilliant red, or yellow, obovate, flattened and hanging, giving rise to the common name of Rocket. Flowering December-January. Widespread in montane grasslands and on mountains. Tas, endemic.
Proteaceae

44 *Gentianella diemensis* Gentian

45 *Helipterum albicans*

46 *Herpolirion novae-zelandiae* Sky Lily

47 *E. collina* ssp *diemenica* Eyebright

48 *Gaultheria hispida* (fruits) Snow Berry

49 *Bellendena montana* Mountain Rocket

50 *Bellendena montana* Mountain Rocket

Podolepis jaceoides
A large flowered yellow daisy with several slender stems arising from a woody rootstock. Leaves linear-lanceolate in a basal rosette, smaller and narrower on the flowering stem. Flowers usually solitary 2-3 cm across, outer bracts at base of head brown, papery and thin, outer florets tubular at base, then deeply cut into several narrow lobes, disc florets numerous, tubular, shortly 5-lobed. Seeds and pappus minutely rough. Flowering December-January. Widespread, sea level to mountain summits, common in montane grasslands. Tas, Vic, NSW, SA.
Asteraceae

Euphrasia gibbsiae ssp **comberi**
Plant short, usually less than 15 cm, but taller when sheltered, forming a dense clump of upright flowering stems. Leaves short, relatively broad, crowded, with 5-7 short, flat lobes. Upper parts of shoots, bracts and calyces with glandular hairs. Flowers white about 1 cm across with a yellow blotch in the throat and several purple lines on the lobes. Flowering late spring to summer. Alpine heaths and among cushion plants on eastern mountains of South West Tasmania Endemic.
Scrophulariaceae

Trachymene humilis *Alpine Trachymene*
A herb with rosette of stalked bluntly lobed leaves arising from a persistent rootstock. Leaf blade elliptical or ovate, entire or with one or two pairs of blunt lobes. Flower buds pink opening into white flowers in simple terminal umbels 1-2 cm across, stem very short, or 10-15 cm long. Each flower about 4 mm across, 5-petalled. Fruit, pairs of flattened, one-seeded fruits. Flowering December-January. Subalpine, in montane grasslands as at Middlesex and Weldborough. Tas, Vic, NSW.
Apiaceae

Lomatia polymorpha
A sweet scented shrub to 4 m high, the flowers large, creamy-white in heavy heads. Leaves vary in shape from linear-oblong with entire margins to elliptical-oblanceolate toothed near the apex, margins ±recurved, underside densely covered with brown felted hairs; flower stalks also furry. Flowers rather fleshy, numerous in dense terminal and axillary racemes at ends of upright shoots. Fruit, a stalked follicle with persistent long style opening into a wide, ovate, flat plate revealing winged seeds. Flowering January. Widespread on mountains and as undershrub in high altitude forests, to sea level in south and west. Tas, endemic.
Proteaceae

Coprosma hirtella *Coffee-berry*
Shrub 1-2 m high with stiff erect branches. Leaves opposite, broadly ovate, shortly stalked, tapering to a narrow point, 2-6 cm long, rather thick and rough on upper surface. Flowers in small clusters, axillary or terminal. Male and female flowers on separate bushes. Fruit a 2-seeded globose orange to red drupe, sometimes black. Flowering January. Frequent on rocky hillsides and foothills of mountains. Tas, Vic, NSW, SA.
Rubiaceae

Velleia montana
A perennial herb with a rosette of oblanceolate thick leaves closely pressed to the ground. Flowers in short stalked cymes which are usually prostrate. Flowers with 3 unequal sepals and a ±tubular yellow corolla, split almost to base on upper side, petal lobes short, 5 stamens, style with cup-shaped end. Fruit a many-seeded capsule. Whole plant 10-15 cm across. Flowering November-December. Locally abundant on wet ground at moderate altitudes on the Central Plateau and upper end of valley of St Pauls River. Tas, Vic, NSW.
Goodeniaceae

51 *Podolepis jaceoides*

52 *Euphrasia gibbsiae* ssp *comberi*

53 *Trachymene humilis* Alpine Trachymene

54 *Lomatia polymorpha*

55 *Coprosma hirtella* (fruits) Coffee-berry

56 *Velleia montana*

RAINFOREST

Forest of the heaviest rainfall areas taking up to 300 years to become fully developed under conditions of consistent humidity and freedom from fire.

Nothofagus cunninghamii *Myrtle, Myrtle Beech*
In moist sheltered conditions this tree grows to 35-50 m with many spreading branches. Its small, flat, leathery leaves with glandular surface and dentate margins form a dense lacy canopy, shading out most plants except ferns. In spring the young leaves are a conspicuous bronze or red colour. The separate male and female flowers are carried on new growth. Male flowers are solitary, stalked, with pendulous stamens. Female flowers with protruding styles are unstalked in threes in axils near the end of branches. The trees are wind pollinated. Fruits are small winged nuts with an outside husk as in the related Beech. Flowering November-December. Wet gullies and rainforest. Tas, Vic.
Fagaceae

Atherosperma moschatum *Sassafras*
A conical tree to 45 m with spreading rather drooping branches. Oil glands in bark and leaves give a distinctive sarsparilla-like smell and taste. Bark grey and white, smooth. Leaves ovate-lanceolate, pointed and toothed, in opposite pairs, green and shining above, yellow-white beneath. One side of the spray thus shows all leaves green, the other side all white. Stalked flowers on the underside of branchlets face downwards. Trees often unisexual, male flowers about 2 cm across with 8 white perianth members and about 12 short thick stamens. Female flowers smaller, less round and less showy with many hairy carpels embedded in the centre, bristly styles protruding. Sometimes both male and female flowers are on the same tree. Fruit dry — a blackish, woody knob with feathery styles protruding and gradually decaying. Bark used for flavouring beverages, timber for craftwork. Flowering September-October. Widespread in rainforests. Tas, Vic, NSW, Qld.
Monimiaceae

Agastachys odorata *White Waratah*
An erect bushy shrub to 3 m with erect branches, or on the west coast may be a small tree. Thick fleshy oblong leaves, 4-5 cm long, without hairs, with short thick stalks and entire margins, often carried in a semi-erect position. The strongly scented flowers are creamy white in straight upright spikes 5-8 cm long, in the leaf axils near the ends of branches. Flower buds, tubular, opening into 4 equal spreading lobes to which the stamens are attached; ovary triangular. Fruit a one-seeded winged nut. Flowering December. Local in areas of high rainfall, in scrub associated with button grass plains. Tas, endemic.
Proteaceae

Eucryphia lucida *Leatherwood*
This fair sized tree of rainforests is the source of nectar for Leatherwood honey. The leathery stalked leaves are opposite, their stipules joined across the stem. Young buds and tips of shoots inside these cups are covered with a yellow gummy substance. Large 3 cm flowers in leaf axils near the ends of branches are so numerous in good seasons that the whole tree is white. The flower has deciduous sepals, 4 orbicular petals, very numerous long stamens with anthers which are red before opening, then purple and fading to grey, and a long erect ovary. Fruit, a thin woody elongated capsule. Flowering December-January. Rainforests of the west, centre and south. Tas, endemic.
Eucryphiaceae

Prionotes cerinthoides *Climbing Heath*
A crimson-flowered climbing shrub of the rainforests of the west and south-west. On ground or epiphytic growing on the trunks of trees, sometimes to a height of 10 m, it has long slender branches bearing glossy dark green leaves, paler beneath, 1.5 cm long, with toothed margins. The pendulous stalked flowers are crimson, tubular, about 2 cm long, constricted at the throat, long stamens bringing the anthers to the mouth of the tube. Style and stigma protruding. Fruit a woody capsule opening by 5 valves. Flowering November-April. Tas, endemic.
Epacridaceae

57 *Nothofagus cunninghamii* (forest)
Myrtle forest

58 *Nothofagus cunninghamii* Myrtle

59 *Atherosperma moschatum* Sassafras

60 *Agastachys odorata* White Waratah

62 *Eucryphia lucida* Leatherwood

63 *Prionotes cerinthoides* Climbing Heath

61 *Eucryphia lucida* Leatherwood

Richea pandanifolia *Giant Grass Tree, 'Pandani'*
A tree to 12 m, usually unbranched, lower trunk bare or clothed with persistent dead leaves. Numerous leaves up to 1.5 m long and 3.5 cm wide form a crown on top of the trunk, each leaf with a wide sheathing base and a tough blade tapering to a long point, the margins coarsely toothed. Red or pink flowers in dense stalked heads in leaf axils. As is typical with *Richea* spp. the caps of petals are shed to reveal the stamens. Flowering November-January. Wet forests of mountains and south-west Tasmania. Tas, endemic.
Epacridaceae

Richea dracophylla
An erect slightly branched shrub to 2 m sometimes to 5 m. Leaves spirally arranged with bases sheathing the stem; blade tapering into a long sharp point, averaging 22 cm overall. Flowers white in large dense terminal spikes 13 cm long, with several large brown leaf-like bracts protruding from the head. Numerous tiny 5-segmented fruits splitting when dry. Flowering September-November. Montane and in rainforests. Tas, endemic.
Epacridaceae

Athrotaxis selaginoides *King Billy Pine*
A tree to 40 m in a favourable situation but often stunted and twisted in high exposed places. The thick leaves to 12 mm long overlap loosely, their tapered points projecting or curving back towards the stem. Seed cones spherical, about 1.5 cm diameter, at the ends of branchlets. A valuable commercial timber, straight grained, durable, pink in colour and easily worked, used for window frames and boat building. Found on valley slopes in high rainfall areas between 800 and 1400 m often in association with *Nothofagus cunninghamii* (Myrtle). Tas, endemic.
Taxodiaceae

Anopterus glandulosus *Native Laurel*
A tall shrub of the high rainfall forests, often under a canopy of Myrtle, *Nothofagus cunninghamii.* In such shaded positions the plant becomes straggly and the branches tend to layer forming thickets. Leaves elliptical, about 12 cm long, stalked, thick and glossy, margins with coarse well spaced teeth. Flowers in terminal racemes are cup shaped, 2 cm across with small sepals, 6 fleshy concave white or pale pink petals, 6 straight stamens and a long upright conical ovary of 2 carpels with short diverging styles. Fruit, a capsule containing 2 winged elongated seeds. Flowering October. Wet eucalypt forests and rainforest from sea level to 1000 m. Tas, endemic.
Escalloniaceae

Athrotaxis cupressoides *Pencil Pine*
An endemic tree of high rainfall montane areas at altitudes between 700 and 1500 m often at edge of streams, tarns and lakes. Tree conical in shape with markedly tapering trunk to 15 m. Foliage, narrow rounded twigs 3-4 mm diameter, covered with closely overlapping stem-hugging leaves, smooth to touch. Male and female cones on same tree. Male cones terminal hardly wider than foliage. Mature fertile cones 1 cm across, spherical, at end of twigs. Same family as the California Redwood, *Sequoia sempervirens.* Tas, endemic.
Taxodiaceae

Phyllocladus aspleniifolius *Celery-top Pine*
A widespread endemic tree to 20 m in wet sclerophyll forests and rainforests from sea level to 900 m. Male flowers insignificant. The female 'flower' consists of a seed-bearing scale which becomes pink and fleshy, the seed surrounded by a white outgrowth from its stalk as illustrated. The so-called leaves (cladodes) are thick, leathery and in shape reminiscent of celery leaves. A valuable commercial timber that does not shrink when dried. Specimens were sent to William Hooker at Kew (UK) by Ronald Campbell Gunn in 1844. Tas, endemic.
Phyllocladaceae

64 *Richea pandanifolia*
Giant Grass Tree, Pandani

65 *Richea pandanifolia* (flowers)
Giant Grass Tree, Pandani

66 *Richea dracophylla*

67 *Athrotaxis selaginoides* (cones)
King Billy Pine

68 *Anopterus glandulosus* Native Laurel

69 *Athrotaxis cupressoides* Pencil Pine

70 *Phyllocladus aspleniifolius*
Celery-top Pine

Trochocarpa gunnii
Tall shrub to 6 m in sheltered areas. Small sweetly scented pinkish-white flowers in short dense spikes near the ends of the branches. Leaves small, ovate, shortly stalked, 6-10 mm long, 3-4 mm wide, in opposite pairs, dark green above with prominent longitudinal veins on the lower surface. Fruit is a globular drupe 6-8 mm diameter, pale to dark blue or occasionally orange. Flowering December-January. Widespread but not common, in rainforests of south, centre and west. Tas, endemic.
Epacridaceae

Lagarostrobos franklinii *Huon Pine*
A tall much-branched tree 20-38 m high, with spreading or drooping branches but rising to a dense pyramidal crown. Foliage angular, leaves very small and thick, decurrent. Male cones ovoid, terminal, inconspicuous, hardly wider than the foliage. Female cones very small and loose, made up of 6-10 well-spaced seed-bearing scales about 4 mm across, each with a single seed. Cones seen in January. In many parts of the rainforest in south-west Tasmania along rivers, on swampy ground and fringing lakes, sea level to 850 m, and from Pieman River to western edge of the Central Plateau and to the middle reaches of the Huon River. Tas, endemic.
Phyllocladaceae

Cenarrhenes nitida *Native Plum*
A spreading shrub or small tree to 10 m in rainforests of the west, south-west and Central Plateau. Leaves up to 12 cm long oblanceolate with a blunt tip, cleanly and decisively toothed; when crushed they have a nauseating smell. Flowers white, nearly 1 cm across in near terminal axillary spikes; each has 4 pointed white lobes, 4 thick stamens, one with a touch sensitive hair on the anther which triggers the pollination mechanism. Fruit, large purple spherical drupe up to 1.5 cm wide. Flowering November-December. Tas, endemic.
Proteaceae

Anodopetalum biglandulosum *Horizontal*
A slender tree with a habit of bending over and sending up vertical branches which in turn bend down, finally forming an impenetrable scrub with a rough platform several metres above the ground. Found in southern and western rainforests with a rainfall in excess of 1750 mm and poorly drained acid soil. Small flowers with 4 pale green pointed petals giving a square effect. Leaves, light green, shining, narrow elliptical with a blunt point and coarsely serrated margins. Fruit about 6 mm, fleshy and green with one seed. Tas, endemic.
Cunoniaceae

Cotula filicula
A very small herbaceous creeping daisy with pinnately divided leaves, each division slightly lobed. Leaves rather thick and fleshy, the whole plant with glandular hairs and some straight hairs. Flowers about 6 mm in diameter, without rays but having minute greenish-yellow tube florets sitting on a conical axis, outer florets female, inner ones with stamens. Seeds small with short wings and without pappus. Flowering January. Mountains and mountain rainforests.Tas, Vic, NSW.
Asteraceae

Persoonia gunnii
An erect bushy shrub or small tree to 4 m, bark grey, scaly; young branches and leaves densely covered with short hairs. Leaves curving upwards, narrow spathulate to 6 cm long, flat and thick, no veins showing. Scented flowers, cream or soft yellow about 15 mm long, solitary in the leaf axils and more numerous near the ends of the branches. Four perianth members with frilled margins, rolling back to give a bell-like appearance. Stamens projecting. Fruit an ovoid drupe becoming purple-black about 1 cm long. Flowering in summer. Wet forests and sub-alpine shrubberies. Tas, endemic.
Proteaceae

71 *Trochocarpa gunnii*

72 *Lagarostrobos franklinii* Huon Pine

73 *Cenarrhenes nitida* Native Plum

74 *Anodopetalum biglandulosum* Horizontal

75 *Cotula filicula*

76 *Persoonia gunnii*

WET SCLEROPHYLL

Forests of high rainfall with a canopy of eucalypts as the tallest trees. This section includes species which grade into wet rainforest on one hand and into dry sclerophyll on the other.

Eucalyptus delegatensis *Gum-topped Stringybark, Mountain White Gum*
A massive tree up to 90 m. Bark on the lower trunk thick, grey, rough, fibrous, upper part of trunk and branches smooth white, streaked with grey. Often one or two branches spread outwards then grow erect, forming 'elbows'. Juvenile leaves stalked, opposite for a few pairs then alternate, broadly elliptical, greyish-green with white bloom; old juvenile leaves characteristically red. Mature leaves alternate 8-16 cm long, 2-5 cm broad, stalked, broadly ovate-falcate, leathery, slightly bluish-green. Flowers white, in groups of 7-15 on thick stalks 10-12 mm long. Capsules stalked, pear-shaped, 10-15 mm in diameter, flat-topped, 4 chambered. An important hardwood tree, marketed as Mountain Ash. Flowering January. Widespread above 450 m forming extensive pure stands in the north-east and on Central Plateau. Tas, Vic, NSW.
Myrtaceae

Pittosporum bicolor *Cheesewood, Tallow-wood*
A small neat tree in wet eucalypt forests but at high altitudes reduced to a stunted shrub. Leaves elliptical, 2-6 cm long, about 6 mm wide, with revolute margins, dull green above and rusty brown underneath with short felted hairs. The flowers are bell-shaped, the petals cohering rather than joined, one or several yellow or red flowers flushed with deep crimson or purple, hanging on stalks from the ends of the branches. Fruit, a rounded ovoid capsule, at first yellowish-green then woody, splitting along two sides and opening out, the seeds in a sticky red pulp. Flowering September-November. Widespread in wet forests and gullies, sea level to 1500 m. Tas, Vic, NSW.
Pittosporaceae

Olearia argophylla *Musk*
Sometimes a tree to 15 m but often a much-branched understorey shrub or small tree. Leaves shortly stalked broadly elliptical-ovate, green above, silvery below due to covering of very short white branched hairs. Daisy-type flowers 1 cm across with only a few white rays and cream centres, very numerous in large terminal heads 16 cm across. Seeds with long straight pappus bristles. Timber when large enough was formerly esteemed for cabinet making, inlays etc. Flowering November-December. Common and widespread in wet eucalypt forests and in fern gullies from sea level to moderate altitudes. Tas, Vic, NSW.
Asteraceae

Telopea truncata *Waratah*
The beauty of this shrub in full flower with its crimson heads is not easily forgotten. A small much-branched tree, or spreading shrub up to 8 m high, growing in mountain forests at altitudes between 600 and 1200 m. Young branches often rust-coloured, covered with soft hairs. Leaves up to 18 cm long, wider towards the tip which may be blunt or sharp. Upper surface of leaves dark green, under surface paler, covered with fine soft hairs. Terminal flower heads 5-8 cm across consisting of 15-20 small, scarlet, deep red or very rarely yellow flowers are unmistakable. Each floret has 4 narrow petals, inconspicuous stamens, and long curved style, and produces copious nectar, which drips from the flower. Fruit is a dark brown woody follicle containing two rows of winged seeds. Flowering November-December. Tas, endemic.
Proteaceae

Bedfordia linearis
Shrub or small tree with slender trunk and thin spreading branches, bark grey, upper twigs covered with white tomentum. Leaves alternate, narrow-linear to 9 cm long, margins revolute, under surfaces white with stellate hairs. Flower heads clear golden yellow, white stalked, one or two in each axil of many leaves near the ends of branches making a showy mass but flowers much shorter than the subtending leaves. Daisy-type head with all florets tubular, head 6 mm across. Phyllaries green with white felted hairs, inner ones with shining scarious margins. Pappus bristles long, white. Flowering December-January. Widespread in wet eucalypt forests and on rocky hillsides. Tas, endemic.
Asteraceae

77 *Eucalyptus delegatensis*
Gum-topped Stringybark, Mountain White Gum

79 *Olearia argophylla* (scene) Musk

80 *Olearia argophylla* Musk

78 *Pittosporum bicolor*
Cheesewood, Tallow-wood

81 *Telopea truncata* Waratah

83 *Bedfordia linearis*

82 *Telopea truncata* Waratah

Cyathodes glauca *Cheeseberry*
A leggy shrub usually to 3 m, with semi-erect or spreading branches arising in rings at intervals from the stem; leaves often persisting just below the branches and crowded in apparent whorls near the ends of them, with intervals of bare stem between. Leaves 1.5-3 cm long, narrow-elliptical, pointed, striate beneath. Flowers 6-10 mm long in terminal axils, white, tubular, with slightly hairy leaves. Berries are flattened drupes, pink, mauve, purple or white, 8 mm in diameter, in leafy clusters at ends of branches. Flowering December-January. Common and widespread on wet hillsides and mountain foothills to 1100 m. Tas, endemic.
Epacridaceae

Monotoca glauca
A densely branched shrub or small tree to 8 m with rather slender branches. Leaves shortly stalked, linear-elliptical, pointed but not sharp, striate and pale underneath. Flowers small, solitary or in axillary spikes; trees often unisexual. Fruit a greenish drupe. Flowering late summer. Common understorey shrub at edges of wet forest and in previously logged areas. Tas, endemic.
Epacridaceae

Phebalium squameum *Lancewood, Satinwood*
Erect shrub or small tree, branches semi-erect; in the north-west may grow to 12 m. Leaves leathery, dotted with translucent oil glands, aromatic, lanceolate or narrow-oblong 4-8 cm long, underside silvery. Flowers in small clusters on thick scaly stalks in axils of leaves; petals 5, white, to 6 mm long, stamens 5. Fruit separating into one-seeded parts. Timber when large enough was formerly used in cabinet work. Subspecies *retusum* has shorter leaves and very warty scaly twigs. Flowering spring. Widespread in wet forests and along river banks. Tas, endemic.
Rutaceae

Zieria arborescens *Stinkwood*
A tall shrub or small tree up to 5 m found in wet forests. Leaves opposite, trifoliolate, dark green with disagreeable and unmistakable smell when crushed. Flowers in loose stalked clusters in leaf axils near ends of branches. Each flower has 4 white or pink broadly elliptical petals. Browsing cattle may be poisoned by eating this plant. Flowering spring. Widespread in wet gullies and forests. Tas, Vic,
Rutaceae

Westringia rubiaefolia
A compact much-branched shrub to 1.5 m with close whorled leaves and white 2-lipped flowers. Leaves sessile in whorls of 4 around a rather square stem, each leaf 5-10 mm long, flat, thick, with prominent mid-rib and slightly recurved margins. Flowers solitary in leaf axils towards ends of branches. Flowers small, open tubular, 2-lipped, upper lip broad and 2-lobed, lower lip 3-lobed and spreading. Inside surfaces softly hairy, white or pale pink spotted with lilac. Sepals 5, 2-3 mm long, rough. Fruit a group of 4 nutlets. The genus *Westringia* can be distinguished from the closely related *Prostanthera* by its whorled leaves and 5-lobed calyx, while *Prostanthera* has opposite leaves and a 2-lobed calyx. Flowering December-April. Widespread in wet eucalypt forests, along creeks and in montane areas. Tas, endemic.
Lamiaceae

Tasmannia lanceolata *Mountain Pepper*
Although the leaves, bark and berries have an aromatic peppery taste, burning the mouth, the berries are eaten by native birds such as the Black Currawong. A compact bushy shrub; young stems and leaf stalks red, leaves elliptical-oblanceolate, hairless, green and thick. Length of leaf varies from 1.5 cm in harsh conditions to 13 cm. Male and female flowers on different plants at the base of the new season's growth. Sepals deciduous, a variable number of yellow or cream narrow oblanceolate petals. Male flowers with many stamens; female with 2-lobed ovary. Fruit 2-lobed lustrous black berry with many small angular seeds. Flowering October-November. Widespread, sea level to subalpine in high rainfall areas and montane grasslands and shrubberies. Tas, Vic, NSW.
Winteraceae

84 *Cyathodes glauca* Cheeseberry

87 *Zieria arborescens* Stinkwood

88 *Westringia rubiaefolia*

89 *Tasmannia lanceolata* (female flowers)
Mountain Pepper

85 *Monotoca glauca*

86 *Phebalium squameum*
Lancewood, Satinwood

90 *Tasmannia lanceolata* (male flowers)
Mountain Pepper

Hakea lissosperma *Needle Bush*

A small tree 2-6 m high with hard cylindrical leaves 3-5 cm long, ending in stiff sharp points. Flowers white, clustered in leaf axils on stalks 4-6 mm long. Fruit a hard woody capsule 2-3 cm long, 1.6-2.2 cm broad, brownish-purple, rather warty with a small often inconspicuous beak. Some trees have abundant flowers but set no fruit, the flowers seem to be functionally male or female with sexes on different trees. Flowering October. Widespread, especially on mountains to 1200 m and to sea level in areas of high rainfall. Tas, Vic, NSW.
Proteaceae

Coprosma nitida *Shining Coprosma, Mountain Currant Bush*

A spiky, much branched erect shrub to 3 m, the smaller branches ending in spines. Leaves opposite, elliptical, flat, thick and glossy 5-20 mm long. Flowers small, unisexual, solitary and terminal on very short axillary shoots. Sexes on separate plants. Male flowers have tiny sepals, longer petals and long pendulous stamens; female flowers have sepals, petals and two long protruding slightly feathery styles. Fruit 2-seeded drupes, ovoid to spherical, translucent, orange, shining, often very numerous, edible but not very palatable. Flowering September-October. Montane from 800 m to mountain summits where it is reduced to a prostrate rock hugging small shrub. Tas, Vic.
Rubiaceae

Olearia glandulosa *Swamp Daisy Bush*

This spicy smelling graceful shrub 1-2 m high has small stalked clean white daisy flowers in flat topped tufts at the ends of branches. The alternate leaves are narrow, cylindrical, fleshy and knobbed with small glandular swellings. Flowering November-February. This shrub is found in soaks and wet places, lowland to montane. Tas, Vic, NSW, SA.
Asteraceae

Australopyron pectinatum *Comb Wheat Grass, Spiked Blue Grass (NZ)*

Rhizomatous tufted short-leaved grass with spiky flower heads 5 cm long. Leaves soft, bluish-green, sheathing bases ridged, blades coarsely hairy with many short stiff white hairs. Flower head of 8-15 sessile well-spaced large spikelets each containing many flowers; each spikelet 1.5 cm overall. Bracts of spikelet minutely hairy, produced into long tapering points. Spikelets standing at right angles to the stem or the lower ones pointing down. Montane grasslands, wet rocky and shady places. Tas, Vic, NSW; introduced into New Zealand.
Gramineae

Blandfordia punicea *Christmas Bells*

A spectacular, robust red-flowered lily up to 1 m tall arising from a thickened rootstock. Leaves 15-45 cm long, narrow-linear, firm, tough, ridged, margins slightly serrate and recurved, leaf bases sheathing. Flowers numerous in a large terminal raceme on a stout erect stem 30 cm-1 m long. Individual flowers brilliant red, bell-shaped, pendulous on short stalks, inside of flower and reflexed tips bright yellow, occasionally whole flower yellow. Stamens just inside the tube but conspicuous, style protruding. Fruit long-stalked lanceolate capsules, splitting into three. Flowering October-March. Wet heaths, moors and hillsides, sea level to subalpine in high rainfall areas of north-west, west and south. Tas, endemic.
Liliaceae

Billardiera longiflora *Mountain Blue Berry*

Slender woody plant with twining stems, climbing over other plants. Leaves alternate, narrow-elliptical. Flowers pendulous, 5-petalled, greenish-yellow or flushed with purple, tubular bell-shaped, 1-3 cm long. Berry ovoid-oblong, purple occasionally white, 2 cm long, fleshy and spongy. A red-berried form which differs slightly in shape of leaf and berry, and length of flower is included. Flowering October-January. Abundant, widespread in wet forests. Tas, Vic, NSW.
Pittosporaceae

91 *Hakea lissosperma* Needle Bush

93 *Olearia glandulosa* Swamp Daisy Bush

92 *Coprosma nitida* (fruits)
Shining Coprosma, Mountain Currant Bush

94 *Australopyron pectinatum* Comb Wheat Grass

95 *Blandfordia punicea* Christmas Bells

96 *Billardiera longiflora*
Mountain Blue Berry

97 *Billardiera longiflora* (fruits)
Mountain Blue Berry

Oxylobium ellipticum *Golden Rosemary*
Spreading much-branched shrub to 2.5 m with smoothly pilose branchlets. Leaves elliptical 1-4 cm long, dark green, leathery, reticulate veins and mid-rib obvious, margins recurved. Leaf apex blunt, often with an abrupt point; lower surface of leaf brownish with short hairs. Golden-yellow pea flowers in dense terminal clusters. Pods 7-8 mm long, rounded, grey-brown covered with long silky hairs. Flowering spring-summer. Widespread from sea level to 1000 m. Tas, Vic, NSW, Qld.
The endemic Australian genus, *Oxylobium*, named by the early botanist Robert Brown has 2 species in Tasmania, *O. ellipticum* (shown here), and *O. arborescens* — a larger shrub or small tree to 7 m, leaves 2-8 cm long, pointed not blunt, and orange-yellow flowers in axillary racemes. It occurs in wet forests and gullies.
Leguminosae-Papilionatae

Pultenaea juniperina *Prickly Beauty*
This short prickly shrub sends up shoots from underground stems and so forms dense thickets. Leaves are crowded, narrow, spreading at right angles to the stem, pointed, concave above, about 1 cm long. Flowers fairly large, almost 1 cm across, solitary or several together, axillary and terminal on short side branches, sometimes very numerous and showy but often sparse, scattered at ends of branches. Petals orange-yellow, keel dark purple-brown. Pod ovate, short and fat. Flowering October-January. Widespread, sea level to mountain plateaux. Tas, Vic, NSW.
Fabaceae-Faboideae

Cyathodes parvifolia *Mountain Berry*
Small dense shrub to 1 m high in wet cold areas. In May the female plants are densely covered with spherical berries about 6 mm across in varying shades of pink and red. Leaves are small about 6 mm long, linear-lanceolate, prickly, dark green, with fine white stripes underneath. Numerous small cream bell-shaped flowers are borne in the leaf axils; male and female flowers on different plants. Flowering September-December. Common on rocky slopes and wet grassy plains up to an altitude of 1200 m. Tas, endemic.
Epacridaceae

Olearia lirata *Dusty Daisy Bush*
A tall shrub or small tree of wet forests with white daisy flowers and dark green leaves with yellowish-grey under-surfaces. Young shoots, leaf stalks and undersides covered with yellow-white stellate hairs, leaves lanceolate, 6-15 cm long, sometimes wrinkled or margins toothed. Daisy heads white, numerous, in large showy clusters at ends of branches, each flower head about 1 cm in diameter. Fruit with pappus bristles 4 mm long. An excellent tree for native gardens. Flowering September-February. Widespread on margins of wet forest and in gullies, even at low altitudes. Tas, Vic, NSW.
Asteraceae

Olearia phlogopappa *Daisy Bush*
A very widespread daisy bush; several distinct varieties are recognised. Leaves and young branches greyish with dense covering of stellate hairs. Upper surface of older leaves usually green (one variety retains the greyish-yellow hairs). Leaves narrow-elliptical, margins crenate or irregularly toothed or occasionally entire. Flower heads white, 2 cm across, numerous in branched inflorescences or solitary, on terminal and lateral shoots. Blue or pink forms exist. Fruit with pappus. Flowering spring. Common in many places from sea level to mountains, frequent on wet hillsides. Tas, Vic, NSW.
Asteraceae

Helipterum anthemoides *Chamomile Sunray*
A perennial everlasting daisy with a woody rootstock from which rise many smooth erect slender unbranched stems, 20-40 cm high. Leaves smooth, blunt, bluish-green, alternate, narrow-linear, 6-15 mm long. Delicate white flower heads with small yellow discs, 15-25 mm wide, outer bracts broad, white, papery and tinged with pale brown. Flowering January-February. Widespread, occasional in lowlands and abundant in montane grassland, e.g. Middlesex Plains. Tas, Vic, NSW, Qld.
Asteraceae

98 *Oxylobium ellipticum* Golden Rosemary

99 *Pultenaea juniperina* Prickly Beauty

100 *Cyathodes parvifolia* Mountain Berry

101 *Olearia lirata* Dusty Daisy Bush

102 *Olearia phlogopappa*
Daisy Bush

103 *Helipterum anthemoides*
Chamomile Sunray

Helichrysum acuminatum *Orange Everlasting*
A perennial herbaceous daisy with branching underground stems forming large flat clumps, bearing crowded leaves and erect leafy flowering stems. Leaves narrow-lanceolate to obovate, bluntly pointed, surface rough, sometimes with cobwebby hairs at base and on margins. Large flower heads 3-4 cm across, with stiff bracts, straw-like in texture, outermost brownish, inner bright golden-yellow, occasionally white, on unbranched stems. Disc florets yellow, pappus of stiff yellow hairs. Flowering January-March. Widespread, montane grasslands and heaths to mountain summits. Tas, Vic, NSW.
Asteraceae

Acacia melanoxylon *Blackwood*
Large erect tree to 30 m with dense crown and dark furrowed bark on trunk. Phyllodes are dull grey-green, ± elliptical, straight or curved 4-10 cm long, 10-25 mm wide with 3-5 prominent longitudinal veins; glands near base. Dense pale yellow heads, globular, of 30-50 flowers on stout stalks 5-10 mm long, solitary or in short racemes of 2-8 heads. Pods flattish, curved or coiled 4-12 cm long, 6-10 mm broad, margins thickened, slightly constricted between seeds. Suckers readily from damaged roots, young leaves bipinnate. A valuable commercial timber which is used extensively for furniture making, panelling and for craftwork. Flowering August-October. Widespread, in wet gullies and forests. Tas, Vic, NSW, Qld, SA.
Fabaceae-Mimosoideae

Prostanthera lasianthos *Christmas Bush, Mountain Lilac*
Small tree to 6 m with dark red bark on upper limbs, branches opposite. Leaves opposite, lanceolate, margins toothed; flowers white or pale lilac spotted with deep purple, all petals joined into a tube then becoming 2-lipped, lower lip with 3 lobes, upper lip shorter and wider, 2-lobed. Fruits small dry nuts, 4 together. Flowering December-January. Margins of streams to sea level, abundant on mountain slopes in openings in wet forest. Tas, Vic, NSW, Qld.
Lamiaceae

Muehlenbeckia gunnii *Macquarie Vine*
This perennial creeping vine straggles over trees and shrubs in wet forests to a height of 10 m. Leaves stalked, hastate or lanceolate, pointed, 2.5-8 cm long. Flowers pale yellow 8-9 mm across, in straight leafy spikes several centimetres long. Fruit slightly succulent. Flowering November-December. Widespread in wet forests, along roadsides and in fern gullies, may be a nuisance blanketing other vegetation. Tas, SA.
Polygonaceae

Senecio linearifolius *Fireweed*
This vigorous quick growing daisy shrub 30 cm-1.5 m high, is found along roads and on hillsides especially in moderate rainfall areas, where it springs up quickly after fires, forming large clumps. It has distinctive large flat clusters of golden flower heads at the ends of the stems. The long narrow, spreading, stalkless leaves are shiny, dark green on the upper surface and sometimes covered with short, soft, cottony hairs on the lower surface. Flowering November-February. Tas, Vic, NSW.
Asteraceae

Acacia pataczekii *Wally's Wattle*
Named after the forester, Wally Pataczek who first found it. Shrub or small slender tree with blue-green foliage and smooth grey bark, suckering readily; young branches angular. Phyllodes blue-green, elliptical, up to 6 cm long, 8-18 mm wide, mid-vein closer to top margin; base of phyllode running down stem; a raised gland near base. Stalked small bright lemon-yellow globular heads of about 15 flowers in long racemes of up to 30 heads. Short wide brown flat pods, purplish when young. Flowering October. Restricted to high country in north-east, at Tower Hill and Roses Tier. Tas, endemic.
Fabaceae-Mimosoideae

104 *Helichrysum acuminatum*
Orange Everlasting

105 *Acacia melanoxylon*
Blackwood

106 *Prostanthera lasianthos*
Christmas Bush, Mountain Lilac

107 *Muehlenbeckia gunnii*
Macquarie Vine

108 *Senecio linearifolius*
Fireweed

109 *Acacia pataczekii*
Wally's Wattle

Veronica formosa
Shrub to 2 m with slender erect branches, scaly where leaves have fallen. Leaves 7-15 mm long, are smooth, lanceolate, spreading or recurved with a broad base ± fused to the stem, in opposite pairs, making four vertical rows. Flowers 4-petalled, pale to lilac-blue, with two long purplish stamens, in loose axillary racemes towards the ends of branches. Fruit, 2-lobed inflated capsules which split from the top and remain on bushes long after seed has been shed. Flowering September-November. Widely distributed on rocky hillsides in the north, and along some river banks in the east; coastal to about 1250 m. Tas, endemic.
Scrophulariaceae

Notelaea ligustrina *Native Olive*
A tall shrub or sturdy tree up to 10 m in height. Leaves opposite, lanceolate or elliptical, 4-6 cm long, bluntly pointed, dull olive green with minute glandular dots. Tiny white 4-petalled flowers in much-branched axillary panicles. Fruit a drupe, 1 cm x 8 mm, green at first, becoming yellow or pink, finally dark purple. Flowering early spring. Widespread along river banks and on moist shady hillsides. Tas, Vic, NSW.
Oleaceae

Echinopogon ovatus *Hedgehog Grass*
A rhizomatous grass with soft narrow leaves, and weak flower stems about 30 cm high, stem leaves short. The flower head, 2 cm long including the bristles, is a terminal ovoid cluster of minutely stalked spikelets. Each spikelet has two short chaffy outer bracts and a rigid inner flowering bract with two short lateral lobes and a long dorsal awn 6 mm long. Whole head is thus very spiky. Common in cool shaded forests. All States.
Poaceae

Drymophila cyanocarpa *Native Solomon's Seal, Turquoise Berry*
Stem simple, sometimes branched in the leafy portion, arising as bare stems from a tuberous rootstock then arching over and bearing alternate leaves and axillary flowers. Leaves lanceolate, thin, in 2 opposite rows, 8 cm long. Flowers 1.5 cm diameter, white, shortly stalked in the axils of the leaves, facing downwards. Perianth of 6 white segments, spreading; 6 stamens with oblong anthers. Fruit a stalked globular or heart-shaped berry, turquoise-blue or paler, to 16 mm across; seeds 8-10, shining, brown. Flowering November-January. Widely distributed in damp shady places in forests. Tas, Vic, NSW.
Liliaceae

Aristotelia peduncularis *Heart Berry*
The dark pink or red heart-shaped fruits 1-1.5 cm in diameter give the popular name to this straggling shrub with slender branches. The leaves are thin textured and opposite with deeply serrate margins, 1.5-7 cm long. White bell-shaped flowers with four 3-lobed petals hang singly from the axils of the previous year's growth. Flowering November-December. Common in moist shady forests and fern gullies especially in the south. Tas, endemic.
Elaeocarpaceae

Sambucus gaudichaudiana *Native Elder Berry*
A woody plant of moist fern gullies. Annual branches 1-2 m, rise from a perennial rootstock. The pinnately compound leaves are 4-10 cm long with coarsely toothed margins and furrowed veins. The flowers are white, cup-shaped with a short tube and 4 spreading, rounded lobes about 10 mm across. Fruit pale yellow to pink, almost translucent, 3 mm long with 3-4 stones. Flowering summer. Found mainly in the north and on King Island. Tas, Vic, NSW, Qld, SA.
Caprifoliaceae

110 *Veronica formosa*

111 *Notelaea ligustrina* Native Olive

112 *Echinopogon ovatus* Hedgehog Grass

113 *Drymophila cyanocarpa* (fruits)
Native Solomon's Seal, Turquoise Berry

114 *Aristotelia peduncularis* (fruits)
Heart Berry

115 *Sambucus gaudichaudiana*
Native Elder Berry

116 *Sambucus gaudichaudiana* (fruits)
Native Elder Berry

WIDESPREAD

This section contains plants which grow under a wide range of conditions and so are found in several types of vegetation, sometimes ranging from sea level to mountain tops.

Diplarrena moraea *White Flag Iris, Butterfly Iris*
A widespread iris of sandy, peaty or rocky places, with fibrous roots and short rhizomes making large tussocks and masses. Leaves 30-60 cm long, narrow, parallel veined, arising from the base of the plant. Flower stalk tall, smooth, with 2-4 erect bracts and 2 large bracts enclosing the flowers. Flowers stalked, emerging in succession from sheath, bilaterally symmetrical, perianth of 3 large segments, upper one slightly larger than two lateral, enclosing 3 shorter narrow inner segments, white and yellow with purple lines, 3 stamens, only 2 functional, and 3-lobed petaloid style. Fruit a capsule. Flowering spring-summer. Tas, Vic, NSW.
Iridaceae

Grevillea australis
A variable much-branched shrub. The form pictured is common on river banks at low altitudes. It has narrow-linear pointed leaves with recurved margins, 5-40 mm long, 1.5-4 mm wide, lower surface covered with short silky hairs. Flowers white with strong sweet smell, in small heads in leaf axils and terminal on short shoots. Fruit small, woody, opening by a slit; seed with a narrow wing. Widespread on river banks throughout State. Some of the several known forms have much wider leaves; one form common on mountains, between 800-1000 m, is a prostrate rock-hugging shrub having lanceolate leaves with recurved margins. Flowering December-March. Tas, Vic, NSW.
Proteaceae

Orthoceras strictum *Horned Orchid*
A green-grey or yellow-brown orchid, with dark markings, well camouflaged. Stem stiff, erect from 30-60 cm high, surrounded by 3-4 narrow, channelled grass-like leaves, leaves grading into stem bracts. Flowers 1-9, spirally arranged around the stem, each base enclosed by a sheathing bract. Upper sepal forming a smooth slaty forward pointing hood covering the column, lateral petals short, lower two sepals long, narrow, spreading horizontally or standing up like horns above the flower. Labellum small, 3-lobed, brownish, middle of tongue yellow, large yellow hump near base. Colour varies from green with grey, purple and yellow, to brown and purple and yellow. Flowering December-January. Not common — usually in wet heaths. Tas, Vic, NSW, SA, Qld; New Zealand.
Orchidaceae

Banksia marginata *Honeysuckle*
A dense bushy shrub or small tree, often flowering when under 1 m but sometimes reaching 9 m. Bark brownish-grey, smooth; leaves 3-10 cm long, narrow, oblong, silvery beneath, margins entire or toothed. Individual flowers small, soft lemon-yellow to golden, crowded into dense spikes up to 10 cm long, 4 cm wide. Cones becoming buff to grey with age, fertile carpels making hard brown bulges on sides of cones. Fruit opening by a slit, seeds winged. Flowering profusely from spring to early winter, it is a significant producer of nectar for birds and insects. Very common in all areas especially heaths and light forests. Tas, Vic, NSW, SA.
Proteaceae

Dianella tasmanica *Blue Berry*
A blue lily with hard linear leaves arising from underground rhizomes, forming dense clumps and tussocks, sometimes covering large areas of wet hillsides. Leaves long ± folded along mid-rib, arranged in two opposite rows, the leaf margins and mid-rib rough with small teeth. Blue flowers 8 mm across are in spreading panicles; each flower has 6 petaloid segments, 6 yellow stamens and central ovary. Base of each yellow stamen is thickened, the anther oblong and yellow. Fruit, a blue-purple ovoid berry. Flowering spring-summer. Widespread and common especially in wetter areas on rocky hillsides, sea level to mountain foothills. Tas, Vic, NSW.
Liliaceae

117 *Diplarrena moraea* (scene)
White Flag Iris, Butterfly Iris

118 *Diplarrena moraea*
White Flag Iris, Butterfly Iris

121 *Banksia marginata* Honeysuckle

119 *Grevillea australis*

122 *Dianella tasmanica* (fruits) Blue Berry

123 *Dianella tasmanica* Blue Berry

120 *Orthoceras strictum* Horned Orchid

Acacia mucronata

An erect shrub or small tree; large well grown trees are common on Flinders and Cape Barren Islands. Phyllodes very variable in size and shape, from narrow-linear to broadly oblanceolate, more than one longitudinal vein, apex blunt or rounded with an abrupt point. Individual pale yellow flowers in loose axillary spikes. Pods straight or slightly curved — to 10 cm. Flowering September-November. Widespread under a variety of conditions; in wet coastal heath, in gullies, along water courses, sea level to 1200 m. Tas, Vic, NSW.
Fabaceae-Mimosoideae

Brachyscome spathulata var. **glabra** *Blue Daisy*

This daisy has a rosette of spathulate toothed leaves arising from a rhizomatous base. Flower stem to 40 cm, slightly ridged, with a solitary terminal flower, very occasionally a second flower branching from near the base. Flower head purple-mauve 3 cm across, eye yellow. The tiny fruits winged, pappus microscopic. Flowering November-December. Grasslands and open ground in forests, sea level to 1200 m. Tas, Vic, NSW, Qld.
Asteraceae

Stackhousia monogyna *Candles*

Named by Labillardiere to honour John Stackhouse (1742-1819) a botanist of Cornwall. A hairless herb with a perennial base and many ascending unbranched leafy stems 30-80 cm high. Spikes of sessile white flowers, pinkish in bud, with 5 narrow sepals and tubular corolla 6 mm long, lobes shorter, blunt. Fruits sessile dividing into 2-3, rough surfaced fruitlets, not angled. Flowering spring-summer. Widespread sea level to Central Plateau. Tas, Vic, NSW, Qld, SA.
Stackhousiaceae

Leptorhynchos squamatus *Scaly Buttons*

A common perennial herbaceous daisy with solitary yellow flower heads on wiry stems to 20 cm high. Leaves narrow, entire, alternate 11-30 mm long, soft in texture, bluish-green, the lower surface white due to numerous small hairs. Flower heads button-shaped, composed of numerous yellow tubular florets surrounded but not overtopped by small regularly spaced overlapping bracts with acute brown tips. Achenes tapering to a short neck. Flowering September-December. Common, sea level to mountains. Tas, Vic, NSW, SA.
Asteraceae

Lomandra longifolia *Sagg*

Formerly known as *Xerotes longifolia.* A harsh tussocky plant of coastal heaths and dry forests. Leaves linear, at base of plant, harsh with cutting edges, ends notched, stems flat about as long as leaves. Flowers small, crowded into spiny inflorescences, yellowish or light straw coloured. Male (illustrated) and female flowers on separate plants. Fruit capsules globose. Flowering November. Very common and widespread especially in dry stony or rocky places. One other species in Tasmania *(L. nana)* is less than 15 cm high with narrow leaves and flowers almost at ground level in very small heads. Tas, Vic, NSW, Qld, SA.
Xanthorrhoeaceae

Correa reflexa *Native Fuchsia*

A small shrub with stiff branches 45 cm to 3 m high. Leaves rough surfaced, ovate to heart-shaped, convex above, sometimes drooping. Branches and underside of leaves covered with rusty stellate hairs. Flowers tubular, bell-shaped, green or sometimes red with yellow or green tips, hanging from leaf axils near the ends of branches. Calyx cup-shaped with 4 small teeth, petals joined in a tube 1.5-3 cm long, tips of petals recurved, stamens longer than tube, 4 with wide bases alternating with 4 narrow ones, style protruding. Fruit splitting into 4 parts. Flowering spring, sparingly at other times. Widespread on rocky hillsides and in coastal heaths. Tas, Vic, NSW, SA, Qld.
Rutaceae

124 *Acacia mucronata*

125 *Brachyscome spathulata* var. *glabra* Blue Daisy

126 *Stackhousia monogyna* Candles

128 *Lomandra longifolia* Sagg

127 *Leptorhynchos squamatus* Scaly Buttons

129 *Correa reflexa* (green) Native Fuchsia

130 *Correa reflexa* (red) Native Fuchsia

Hibbertia procumbens

A small prostrate woody plant, stems spreading over the ground. Flowers large to 2.5 cm across, golden yellow, petals five, broadly obovate, sepals broad, light brown. Leaves linear, blunt and flat to 2.5 cm long, varying greatly in width, both narrow and wide leaved forms exist. Stamens surrounding the ovary. Flowering spring and summer. Abundant in heaths, sea level to mountain plateaux. Tas, Vic.
Dilleniaceae

Luzula sp. *Wood-rush*

A small rush-like plant with flat more or less hairy leaves. Flower heads terminal, stalked either solitary or several in a loose umbel according to species. Heads globose or ovoid in shades of brown and cream. Each flower head consists of many flowers each surrounded by papery bracts, each flower with a perianth of 6 papery segments, 3 or 6 stamens and 1-celled ovary. Fruit a small nut. No specific name is given for the plant illustrated. *Luzula* differs from rushes *Juncus* spp. in having hairy leaves and one chambered 3-seeded fruits. The genus is under revision and the number of species in Tasmania is greater than previously thought. Various common species exist throughout the State. Most species flowering in spring-summer. Throughout Australia.
Juncaceae

Themeda triandra *Kangaroo Grass*

A large perennial grass with rather coarse but narrow foliage. Leaves are folded or keeled along the mid-rib, pale green when fresh, becoming rusty red when dry. Flowering heads to 1 m high consist of several stalked clusters of green and purple spikelets with very long narrow purple bracts below them. Each cluster contains many spikelets with chaffy bracts, but only one which produces a grain. Fertile spikelets have long bent twisting awns. The whole flower head turns rusty brown when ripe. A native grass, it does not stand heavy grazing but may still be seen in large patches along roadsides and in openings in the forest where the stocking rate is low. Flowering October-November. Sea level to montane. All States (including Central Australia region); New Guinea, Philippines, Asia and introduced into New Zealand.
Poaceae

Thelymitra ixioides *Spotted Sun Orchid*

Slender to robust orchid, 30-90 cm high. Leaf long, 15 cm or more, narrow and channelled. Flowers 1-9 on slender stalks, usually blue or mauve, dotted with deeper blue, outside paler or pinkish, occasionally unspotted; column 3-lobed with white, pink or pale mauve hair tufts on side lobes. Flowering September-November. Widespread, coastal heaths to forested foothills. All States; New Zealand and New Caledonia.
Orchidaceae

Drosera peltata ssp **auriculata** *Sundew*

A slender green insectivorous plant growing from a buried tuber. Lower leaves in rosette or reduced to scales. Stem erect, stem leaves stalked, peltate, shield shaped, covered with sticky hairs longer at the upper pointed 'ears'. The hairs trap and digest small insects. Flowers 1-1.5 cm across, few, terminal, white or pinkish, sepals hairless, smooth, shining. Flowering spring-summer. Common, widespread in heaths and on dry hillsides at low altitudes. Tas, Vic, NSW, Qld, SA; New Zealand.
Droseraceae

Kennedia prostrata *Running Postman*

Plant prostrate, flowers crimson, pea-shaped, 1.5 to 2 cm across. Stems trailing with trifoliolate leaves, each leaflet almost orbicular; pods furry 4-5 cm x 7 mm. Common in coastal areas and light forests, colonising bare ground after disturbances and fire and persisting under bracken. *K. prostrata* is the only Tasmanian representative of the large Australian genus *Kennedia*. Flowering September-November. All States.
Fabaceae-Faboideae

131 *Hibbertia procumbens*

132 *Luzula* sp. Wood-rush

133 *Themeda triandra* Kangaroo Grass

134 *Thelymitra ixioides* Spotted Sun Orchid

135 *Drosera peltata* ssp *auriculata* Sundew

136 *Drosera peltata* ssp *auriculata* Sundew

137 *Kennedia prostrata* Running Postman

Centaurium spp. *Centaury*

Slender, herbaceous plants, 30-40 cm high with entire opposite narrow-ovate leaves about 2 cm long, stem-clasping and forming basal rosettes which persist or have withered by flowering time in different species. Flowers pink, tubular, 5-petalled in flat cymes, followed by delicate 2-valved seed capsules. Flowering spring-summer. In coastal heaths and grasslands, widespread. One native and two introduced species in Tasmania. The genus is widespread in temperate and subtropical regions.
Gentianacae

Epacris impressa *Common Heath*

This common small shrub has short lanceolate sharply pointed leaves. Flowering stems are often unbranched with many flowers singly in leaf axils along much of its length, or with short tufted branches with flowers in short dense heads at the ends. Flowers vary from white through shades of pink to deep red, often growing in large patches of one colour. The species can be identified by 5 small depressions at the base of the flower tube. Flowering throughout the year in different localities. Common and widespread, sea level to 800 m. Tas, Vic, NSW, SA.
Epacridaceae

Gompholobium huegelii *Bladder Pea*

A slender pea 15-30 cm high with a woody base and spreading or ascending branches. Stems and leaves glabrous, often pale green ±glaucous, leaves trifoliolate with narrow lobes. Flowers terminal, solitary or two together on long slender stalks usually creamy yellow, shaded with black on the outside. Pod ovoid. This form is common in sandy heaths but a form with bright yellow flowers and blue-green leaves occurs also, known only on ironstone gravel soils. *Gompholobium* is recognised by its fat pod and black shading on the back of the petals. Flowering November. Widespread in sandy heathland and light forest. Tas, Vic, NSW.
Fabaceae-Faboideae

Boronia pilosa *Boronia*

Pleasantly aromatic shrub to 1 m high, common in moist situations and heaths. Leaves are pinnate with 3-7 narrow leaflets up to 15 mm long, somewhat hairy. Flower has 4 short sepals, 4 ovate pink or white petals, 8 hairy stamens which curve inwards over the ovary. Flowers form showy clusters at ends of main and lateral branches; plant often flowers profusely. Fruit, a group of up to 4 carpels, dry. Flowering September-December. Tas, Vic, NSW, SA.
Rutaceae

Drosera macrantha *Climbing Sundew*

A slender trailing reddish-green insectivorous plant growing from a buried tuber. Stem up to 60 cm long, straggling through undergrowth, internodes long. Upper leaves 5 mm across, long-stalked, often three together, blades orbicular-peltate, cupped and concave downwards, covered with long glandular hairs, producing a sticky 'glue' to trap insects. Flowers several in terminal clusters, white, 1.5-2 cm across, sepals hairy, stamens 5, styles repeatedly branched. Fruit a capsule, seeds black. Flowering October. Common in coastal areas. Tas, Vic, NSW, SA, WA.
Droseraceae

Leucopogon collinus *White-beard Heath*

A small much-branched shrub to 70 cm. Leaves 4-12 mm long, linear, tapering to a blunt point, thin, flat or with recurved margins. Flowers white, either in short many-flowered axillary spikes along short lengths of stem, or in axillary spikes on short lateral branches making short heads. Petals bearded, floral tube about as long as calyx. Several differing forms exist, including a miniature montane form and *L. collinus* var. *ciliatus*, also montane, which has shorter, flatter, ovate-lanceolate leaves, and relatively few flowers in small spikes in terminal axils. Flowering spring-summer. Widespread, abundant in heaths and dry places, sea level to montane. Tas, Vic, NSW, SA.
Epacridaceae

138 *Centaurium* sp. Centaury

139 *Epacris impressa* Common Heath

140 *Gompholobium huegelii* Bladder Pea

141 *Boronia pilosa* Boronia

142 *Drosera macrantha* Climbing Sundew

143 *Leucopogon collinus* White-beard Heath

Pterostylis nutans *Nodding Greenhood*

Stalked leaves 4-5 in basal rosette, blunt with crisped margins. Flowers solitary, bent forward, green, sometimes tipped with brown. Hood wide, inflated and rounded on the back. Lower sepals forming lip with narrow gap, and drawn into rather short points embracing the hood. Labellum pointed, curved, with thickened central rib. Flowering July-November. Widespread in light forest and sandy coastal areas. Tas, Vic, NSW, Qld, SA; also New Zealand.
Orchidaceae

Stylidium graminifolium *Trigger Plant*

A perennial herb with semi-erect narrow-linear leaves forming a basal tuft, from which arise erect flower stems often more than twice as long as the leaves bearing many pink, occasionally white, 4-petalled flowers, the fifth petal very small. Flower stalk and sepals covered with glandular hairs. The whole genus *Stylidium* is notable for its jointed hammer-like column, formed from joined stamens, style and stigma. Insects landing on the flower probe deep into it with their tongues, triggering the movement of the column which transfers pollen to their backs or receives pollen, so ensuring cross pollination. The trigger and column reset slowly. Flowering September-December. Widespread from sea level to montane. Tas, Vic, NSW, Qld.
Stylidiaceae

Microseris lanceolata *Native Dandelion*

Slender yellow-flowered dandelion with a rosette of linear-lanceolate hairless leaves 10-25 cm long, margin entire or with distant small teeth. Flower heads, 2.5 cm diameter, solitary on long stalks, outer bracts herbaceous, green, often with darker edges, without hairs. All florets strap-shaped. Buds nodding. The seeds with pappus of flattened plumose scales can be seen in the illustration. Flowering spring-summer. All temperate States; New Zealand. The only other species in this genus occurs in South America.
Compositae

Comesperma retusum *Purple Milkwort*

A small shrub up to 1 m high, the upper part with several slender branches. Leaves elliptical-ovate about 1 cm long, flat, rather thick, veins indistinct. Flowers bright pink-purple, rarely lavender coloured, 2 large inner sepals forming the wings, other floral parts small. Fruit a wedge-shaped capsule, seeds with a tuft of hairs. This plant belongs to the same genus as *Comesperma volubile*, the Blue Love creeper. Flowering November-December. Widespread in wet heaths and scrub, especially inland. Tas, Vic, NSW, Qld.
Polygalaceae

144 *Pterostylis nutans*
Nodding Greenhood

145 *Stylidium graminifolium* (scene)
Trigger Plant

147 *Microseris lanceolata*
Native Dandelion

146 *Stylidium graminifolium*
Trigger Plant

148 *Comesperma retusum*
Purple Milkwort

Caladenia lyalli *Alpine Caladenia*

Slender or stout plant 8-20 cm high, stem erect, reddish, slightly hairy with single basal ±hairy leaf shorter than stem. Flowers 1, 2 or 3, with broad white rather fleshy segments, outer surfaces bright rosy pink, sometimes paler. Dorsal sepal broad, arching over, hiding the column, other sepals and petals spreading forward. Labellum shortly toothed, marked with red, often red at tip, with 4 rows of yellow or whitish projections (calli). Flowering late spring-summer. Widespread on mountain plateaux, also reaching sea level. Tas, Vic, NSW; New Zealand.
Orchidaceae

Acacia verticillata *Prickly Moses*

A wattle with stalked ovoid heads, widespread especially in damp areas, gullies and along creeks. Usually a straggling bush up to 3 m but a small tree in favourable conditions. The arrangement of the spiky phyllodes (apparent leaves) in rings gives the name *verticillata.* These vary in width from 1-4 mm in different forms. The arrangement of the individual flowers to form a cylindrical spike can be seen in the buds pictured. Flowering September-November. Widespread in damp areas. Tas, Vic, NSW, SA.
Fabaceae-Mimosoideae

Helichrysum thyrsoideum

Small shrub with slender spreading branches. Flower heads numerous in terminal panicles on many erect lateral shoots arising from longer horizontal branchlets forming flat one-sided sprays. Flower heads small, of 6-8 florets surrounded by outer pale brown bracts and inner white-tipped papery bracts; each head about 5 mm long. Seeds with pappus of fine bristles with thickened tips. Flowering January. Common as a regrowth shrub in burnt or logged areas of medium high rainfall; sea level to montane. Tas, Vic, NSW.
Asteraceae

Helichrysum dendroideum

A branched aromatic shrub or small tree with linear-lanceolate leaves to 8 cm long, white or yellowish underneath, except on the mid-rib and very narrowly recurved margins. Branches ribbed. Flowers white in large branched panicles terminal on main and lateral branches. Each flower head small, of about 6 florets ± 3 mm long, outer bracts brownish, inner bracts white tipped. Seeds with pappus. Flowering December. Moist gullies and along creeks — wet margins of forest. Tas, Vic, NSW, SA.
Asteraceae

149 *Caladenia lyallii* Alpine Caladenia

150 *Acacia verticillata* Prickly Moses

151 *Helichrysum thyrsoideum*

152 *Helichrysum dendroideum*

RIVER BANKS AND WET PLACES

This section contains plants which are found in wet places along creeks, river banks and areas subject to occasional flooding.

Sprengelia incarnata *Sprengelia, Pink Swamp Heath*
A short shrub to 80 cm, occasionally taller to 2 m, with erect branches, and hard leaves with sharp points. Leaves alternate, lanceolate with sheathing base completely surrounding the stem, blade curved back tapering to a long point. Older stems smooth reddish brown without leaf scars. Flowers solitary, terminal on short shoots and lateral branches, 5 sepals, greenish pink or white; 5 petals pink or white, very narrow and pointed, herbaceous and widely spreading, 5 stamens spreading or forming a tube in centre of flower, style long. Fruit dry. Whole plant except for petals and stamens more or less hard. Flowering spring-summer. Heaths, wet places, sea level to mountains. Tas, Vic, NSW, SA.
Epacridaceae

Callistemon pallidus *Yellow Bottlebrush*
A bush or small tree with long yellow bottlebrush spikes 4-8 cm long, 3 cm wide. Leaves narrow-elliptical 3-8 cm long, dull green dotted with oil glands just visible to the naked eye. New leaf growth silky hairy, often pink with silvery hairs, hairs lost with age. Sepals and petals of individual flowers pale yellow, under 5 mm, stamens free about 20 per flower, 1.5 cm long, pale yellow. Flowering December. Widespread along streams, on hillsides in wet areas. Tas, Vic, NSW.
Myrtaceae

Leptospermum lanigerum *Woolly Tea-tree*
A much branched shrub or small dense tree, young buds covered with shining straw-coloured bracts, young leaves and shoots covered with soft silky hair. Leaves flat, oblong or ovate, margins slightly recurved, 1-1.5 cm long, silky at least on lower surface. Flowers 1.5 cm in diameter, solitary but numerous, petals white. Sepals and young capsule silky hairy, remaining silky until the second season. Capsule domed, opening by 5 slits. Leaf size variable, one form with leaves 5-8 mm is widespread. Flowering spring-summer. Common, widespread in damp places, river banks, montane grasslands and rainforests of west coast where it may become a tree to 18 m. Tas, Vic, NSW, Qld, SA.
Myrtaceae

Bauera rubioides *Bauera*
A wiry shrub with long spreading branches scrambling over other vegetation and making a dense tangle in wet heaths. Leaves in opposite pairs, sessile, each divided into three lobes so that there appear to be 6 leaflets in a ring. Flowers 1.5-2 cm across, on slender stalks, axillary near the ends of branches; sepals long, reddish; petals 5 or 6 or more, thin textured, pink or white; stamens very numerous, with long filaments, short yellow anthers. Fruit capsular. Flowering spring-summer. Tas, Vic, NSW, SA, WA.
Cunoniaceae

Micrantheum hexandrum
A small shrub with many slender erect branches 1-3 m high; flowers small, numerous, male and female on different plants. Leaves thin, leathery, oblanceolate, pointed, in alternate groups of three. Flowers yellowish or cream, 1-3 in leaf axils. Male flowers 6 mm across, shortly stalked, numerous and thus showy, with conspicuous perianth of 3 small and 3 large members; 6 stamens and rudimentary ovary. Female flowers with perianth, no stamens, long triangular ovary with 3 small spreading styles. Fruit, inflated woody capsule 8 mm long, splitting into 3. Flowering spring. Common on river banks, along water courses and in gullies in east and north. Tas, Vic, NSW.
Euphorbiaceae

153 *Sprengelia incarnata*
Sprengelia, Pink Swamp Heath

154 *Callistemon pallidus* Yellow Bottlebrush

157 *Micrantheum hexandrum*

155 *Leptospermum lanigerum*
Woolly Tea-tree

156 *Bauera rubioides* Bauera

Callitris oblonga *South Esk Pine*

Conifer 2-4 m high, with angled bluish green foliage, the individual leaves minute in rings of three. Pollen cones about 3 mm long, terminal on foliage. Young female cones in stalked clusters near base of branchlets. Mature cones usually clustered, grey, ovoid but pointed, smooth except for a small abrupt protuberance near the apex of each woody scale. Cones of two rings of 3 scales, inner scales about twice as long as outer, opening to shed numerous angular 1-3 winged seeds and much free resin. Local on banks of St Pauls, Apsley and South Esk rivers. Tas, NSW.
Cupressaceae

Epacris exserta

A much branched, erect shrub 60-100 cm high, often densely covered with short spikes of flowers on the main and many lateral branches. Leaves flat, elliptical-lanceolate, narrow, erect or spreading, 5-7 cm long. Flowers white, tubular, rather large, solitary, axillary, usually in short heads but sometimes in long spikes along the length of the stem, anthers and stigma protruding from the tube. Bracts and sepals blunt, often overlapping in rows, straw coloured. Flowering September-November. Local in north, on banks of several rivers including the South Esk. Tas, endemic.
Epacridaceae

Spyridium ulicinum

A bushy shrub, 1-3 m high with many ascending branches; leaves varying in form from pointed narrow-linear, through oblong to cuneate or bifid on different plants. Leaves about 1 cm long, margins revolute, underside with dense covering of very short hairs, upper surface dark green, shining. Flowers white, surrounded by brown bracts, numerous, several together near ends of short lateral branches, sweetly scented. Flowering September-December. Local in several widespread situations in north and south on rocky hillsides and along river banks. Tas, endemic.
Rhamnaceae

Carex fascicularis *Tassel Sedge*

This sedge with its drooping stalked heads has flat hard grass-like leaves to 60 cm, rough with microscopic teeth on the mid-rib and margins. The parallel veins on the underside of the leaves have minute nodules. The triangular stem to 60 cm, bears flowers clustered into long drooping heads, the lower heads are female containing seeds, the upper narrow, tan-coloured ones are male with stamens, but occasionally have a few female flowers at the top. Each fruit is enclosed in a green ribbed sheath with a long beak and small terminal teeth. Common along rivers and in open swampy ground. All States; New Zealand, New Guinea, Java.
Cyperaceae

Hakea nodosa *Yellow Hakea*

A bushy shrub of damp places with slender branches up to 4 m tall, often flowering when about 60 cm high. Leaves green, flexible, cylindrical or somewhat flattened, fine-pointed but not sharp, up to 4 cm long. Flowers tiny, yellow, crowded along upper branchlets, tube 2-3 mm long, slender, splitting into 4 lobes, perianth without hairs. Fruit, brown woody capsule, 2-3 cm long, 1.5-2 cm broad with rough warty surface and a small beak. Seed winged. Flowering May-June. Locally frequent in the north-east, and on Flinders, Cape Barren and Clarke islands. Tas, Vic, SA.
Proteaceae

Prostanthera rotundifolia *Round-leaved Mint-bush*

A purple-flowered bush or small tree about 2 m high, strongly aromatic. Small dull green leaves, shortly stalked, round or fan-shaped, ± folded about the mid-rib 4-10 mm broad. Flowers lilac to purple, petals joined into short cup-shaped corolla, in showy sprays. Flowering September-October. Locally frequent in north and east, along river banks and on rocky hillsides. Tas, Vic, NSW, SA.
Lamiaceae

158 *Callitris oblonga* (fruits) South Esk Pine

159 *Epacris exserta*

160 *Spyridium ulicinum*

161 *Carex fascicularis* Tassel Sedge

162 *Hakea nodosa*
Yellow Hakea

163 *Prostanthera rotundifolia*
Round-leaved Mint-bush

Juncus pallidus
A large rush, usually tall and stout. Leaves reduced to loose basal sheaths, stems smooth, erect, faintly striate, to nearly 1 cm across, 1-2 m high. Pith inside stem continuous, loose, spongy and compressible. Flowers pale, very numerous in loose clusters on stiff stalks, basal floral bract appearing to be a continuation of the stem and the flowers lateral to it. Flowers with 6 perianth members, 6 stamens, central ovary. Ripe fruits ovoid, longer than the perianth. Flowering November-December. Wet places, ponds and ditches, usually lowland or sandy soil. All States; New Zealand.
Juncaceae

Xyris operculata
The leaves of this yellow-flowered plant are slender and erect, usually solid like a rush; flower heads are solitary on leafless stems to 60 cm high. Each ovoid or globose head is formed of closely packed dark brown bracts in 5 vertical rows with several individual flowers buried in the mass. These have a buried tube and 3 free exposed petal lobes, golden yellow and spreading. There are 3 stamens and 3 more reduced to hairy tufts. The ovary is buried in the head, with branched style protruding. Flowers open in succession, usually only one or occasionally two are fully open at one time. At present 4 species are recognised in Tasmania. They differ in the form of the leaf and the flower head. Flowering November-January. Wet peaty heaths, sea level to montane. Tas, Vic, NSW, Qld, SA.
Xyridaceae

Utricularia dichotoma *Bladderwort, Fairies' Aprons*
One, two or more deep purple flowers with 2-lobed perianth on a long bare stalk, occasionally white. Upper lobe small, erect; lower lobe spreading, fan-shaped 1.5 cm across, horizontal, purple with 3 short raised yellow lines in the centre of upper edge; a short cylindrical purple spur points down below flower. Leaves 2 or 3 together at ground level, often on slender rhizomes, thin, linear or spoon shaped, 1 cm long; may be absent at flowering time. Fruit a globular capsule. This plant traps and digests minute aquatic or soil animals such as water fleas and insect larvae in small bladders attached to thin stems lying at or below the surface of the soil. Flowering December-January. Peaty areas, sea level to mountain plateaux in waterlogged soil. Tas, Vic, NSW, Qld, SA.
Lentibulariaceae

Epacris obtusifolia
An easily recognised Heath. Flowers white or cream in spikes at ends of slender often unbranched stems, 30 cm to 1 m. Leaves bright green, ovate-lanceolate, apex blunt, ridged on lower surface, usually appressed to stem. Flowers solitary in leaf axils and forming long spikes, corolla tube longer than calyx. Flowering summer. Damp heaths and wet places. Tas, Vic, NSW, SA, Qld.
Epacridaceae

Melaleuca gibbosa *Small-leaved Melaleuca*
A shrub 1-2 m, dense and bushy with numerous slender branches. Leaves pale to grey-green, small 2-6 mm long, opposite, overlapping in 4 rows, rather thick. Flowers mauve, sessile in small dense oblong spikes 1.5 cm long, with conspicuous long stamens arranged in 5 groups opposite the petals. Fruits woody, sessile, in cylindrical spikes, the branches continue to grow so that old fruits encircle the stem. Seeds very fine, like short hairs. Flowering sparsely throughout the year; main flowering in spring. Wet peaty ground in east, north-east and north-west. Tas, Vic, SA.
Myrtaceae

Melaleuca squamea *Swamp Melaleuca*
A shrub or small tree 1-3 m high with stiff semi-erect branches. Leaves 4-8 mm long, lanceolate, alternate and crowded, with long soft points and apex slightly curved towards stem. Flowers usually pink-purple, less commonly white or yellow, without stalks, in terminal spherical clusters approximately 15 mm across. Fruit, woody capsules persistent for several years. Flowering October-February. Widespread in wet heaths, sea level to approximately 1500 m. Tas, Vic, NSW, SA.
Myrtaceae

164 *Juncus pallidus*

165 *Xyris operculata*

166 *Utricularia dichotoma*
Bladderwort, Fairies' Aprons

168 *Melaleuca gibbosa*
Small-leaved Melaleuca

167 *Epacris obtusifolia*

169 *Melaleuca squamea* Swamp Melaleuca

Drosera binata *Forked Sundew*
Red perennial herb with fibrous roots and underground runners. Leaves few in basal rosette, long stalked, leaf blade reddish, forked, with long sticky hairs which trap insects. Main flower stalk long, bearing clusters of stalked white sweetly scented flowers, 2 cm across. Flowering late spring and summer. Bogs and soaks in wet heaths, sea level to montane. Tas, Vic, NSW, SA; New Zealand.
Droseraceae

Isotoma fluviatilis
A small creeping prostrate perennial herb of swamps and wet areas. Leaves alternate, 4-8 mm long, 2-8 mm wide, green, hairless, elliptical to broadly ovate, margins slightly toothed. Flowers solitary, irregular in shape, on long erect slender stems held well above the leaves, petals blue or white, joined in split tube, lobes spreading, free. Lower petals with darker blue or purple blotch near mouth of tube, anthers joined in a ring around the style. Pollen is shed into this cup-like ring then dispersed when two hairs on the anthers are depressed. Flowering spring-summer. Common in wet places of midlands and north. Widespread in eastern States.
Lobeliaceae

Scaevola hookeri *Creeping Fan Flower*
Creeping perennial herb rooting at the nodes and forming mats. Leaves bluntly ovate-oblong, margins wavy or irregularly toothed, leaf surface rough, ±hairy. Shortly stalked white flowers, red or brown on backs of petals, with fan-shaped corolla of 5 joined petals, open to base on upper side. Stamens free, stigma surrounded by a flattened cup into which pollen is shed. Fruit slightly fleshy, white. Flowering December-March. Widespread in wet heaths, sea level to montane and on margins of coastal lagoons and salt marshes. Tas, Vic, NSW.
Goodeniaceae

Villarsia exaltata
A swamp plant with thick roots and broadly ovate to orbicular leaves, the length of leaf stalk dependent on depth of water in which the plant is growing, stalks longer in deeper water. Leaves radical and on stem, often floating, 4-8 cm long. Flowers yellow 2-3 cm across in branched inflorescences held above water level, sepals ± 8 mm, petals about twice as long, joined at the base, densely hairy at the base of the lobe, margins of lobes not fringed. Stamens conspicuous. Plant spreading by runners as does strawberry. Fruit a capsule. Flowering in summer. Common in swamps and on river flats, sea level to Central Plateau. Tas, Vic, NSW, Qld, SA.
Gentianaceae

Pachycornia arbuscula
This plate shows typical salt marsh vegetation, *Pachycornia* is the large central plant. It is an erect branched shrub growing to 120 cm. Branches green or reddish; stem covered and fused with pairs of fleshy leaves, so appearing segmented. Flowering branches terminal and lateral but similar to vegetative shoots. Minute flowers arise at upper projecting edge of several (2-6) segments. Flowers often 3 together, each having either 1 stamen, or 1 stamen and 1 carpel with 2 minute styles. Flowering October-November. Locally abundant in salt marshes and on low rocky shingly islands occasionally inundated. More common in the south. Tas, Vic, SA.
Chenopodiaceae

Mimulus repens *Monkey Flower*
A small herbaceous hairless plant with small purple widely tubular flowers. Leaves opposite, ovate, small and soft, 2-10 mm long; stem creeping and rooting forming mats. Flowers with very short tube and 5 wide spreading lobes, purple or mauve, inside of throat hairy, yellow and white spotted with red. Fruit capsular. Flowering spring-summer. Wet ground, brackish marshes and shores of saline lagoons. All States; New Zealand.
Scrophulariaceae

170 *Drosera binata* Forked Sundew

171 *Drosera binata* (scene) Forked Sundew

172 *Isotoma fluviatilis*

173 *Scaevola hookeri* Creeping Fan Flower

174 *Villarsia exaltata*

175 *Pachycornia arbuscula*

176 *Mimulus repens* Monkey Flower

Gahnia grandis *Cutting Grass*

A very large tussocky plant with plume-like flowering heads to 3.5 m overall. Leaves grasslike, harsh with sharp cutting edges, 1.5 m long, bases sheathing, lower part of leaf erect then bending over. Flower heads bright brown when young and at flowering time, blackish in fruit. Flowers of dark brown scales with yellow stamens hanging out, short styles. Fruit bright red, hanging from the heads by long black staminal filaments. Bracts of spikelet blunt, spikelets rounded. Flowering late spring-summer, depending on altitude. Common in wet or poorly drained situations, sea level to mountains. Tas, NSW, Qld.
Cyperaceae

Calochilus robertsonii *Red Beard Orchid*

This bearded orchid is common in coastal areas and light forest. The plant stands up to 45 cm high with one narrow, channelled leaf about 25 cm long, and 2 to 9 flowers not all open at the same time. The flowers have a greenish perianth marked with purple lines. Dorsal sepal forming a short wide hood, lateral sepals narrower, spreading on either side of labellum. Labellum base covered with brownish-purple oblong raised glands, which become longer giving way to long purplish-red hairs covering the whole labellum except for ±2 mm at the tip. Occasionally a yellow form of this orchid is found. Flowering late spring to early summer. Widespread. All States; New Zealand.
Orchidaceae

Calochilus paludosus *Strap-bearded Orchid*

Slender bearded orchid with erect flowering stem 30-40 cm high. Narrow, erect leaf, channelled above, shorter than stem, one or two sheathing bracts on stem, a smaller bract below each flower. Flowers 1-4, occasionally more, in a terminal raceme, opening in succession, each fully open for only 1-2 days. Hood and spreading side petals green with red or purple markings. The labellum, 20-25 mm long, has crowded short red hair-like glands at the base which grade into the long bright reddish-brown hairs of the beard. The beard covers most of the labellum except for a strap-like tip up to 10 mm long. Flowering late spring to early summer. Widespread in peaty heaths, also in parts of dry woodlands where water lies in winter. Tas, Vic, NSW, Qld, SA; New Zealand.
Orchidaceae

Epacris lanuginosa *Swamp Heath*

Small woody plant usually branched from the base with slender erect branches to 1 m. Stem with short soft hairs. Leaves lanceolate, spirally arranged, 6 mm long, 2 mm wide, with fine slightly incurved points. Flowers white, tube as long as calyx, rather long in proportion to the spreading lobes; bracts and sepals straw-coloured or brown, papery. Anthers small, dark, at the throat of floral tube; stigma protruding. Flowering spring-summer. Widespread, abundant in wet heaths, sea level to 1000 m. Tas, Vic, NSW.
Epacridaceae

177 *Gahnia grandis* Cutting Grass

178 *Calochilus robertsonii* Red Beard Orchid

179 *Calochilus paludosus*
Strap-bearded Orchid

180 *Epacris lanuginosa*
Swamp Heath

DRY SCLEROPHYLL

In general, plants which are found in areas with less than about 1000 mm rainfall per annum, grading into wet sclerophyll with increasing rainfall or into dry open grassland or heathland.

Eucalyptus pauciflora *Cabbage Gum, Weeping Gum*
Medium sized tree to 20 m, smooth barked except at extreme base, bark streaked with grey and white, branches sometimes drooping. Juvenile leaves large to 14 cm long, 3-8 cm wide, ovate, pointed, grey-glaucous. Adult leaves green, thick and leathery to 15 cm long, 2-4 cm wide, the veins diverging at a small angle from the mid-rib showing several parallel almost longitudinal veins. Flowers 5-15 together in umbels. Fruit capsule about 1 cm across, truncate, pear-shaped or hemispherical, top flat, opening by 3 or 4 pores. Flowering usually in summer. Widespread in the east, north and midlands in rain-shadow areas, but also coastal and on the southern edge of the Central Plateau in well-drained situations. Tas, Vic, NSW, Qld.
Myrtaceae

Rubus parvifolius *Native Raspberry*
A slender straggling plant with erect or spreading ± hairy branches, armed with hooked prickles; compound leaves of 3-5 leaflets with lobed margins, upper surface green, veins impressed; lower surface white. Flowers in short panicles or solitary in axils of upper leaves; five round reddish-pink petals, shorter than the hairy lanceolate sepals. Fruit small clusters of 4-10 orange-red drupelets, edible. Flowering December-January. Common on rocky hillsides and near coasts. Tas, Vic, NSW, Qld, SA; China, Japan.
Rosaceae

Leptomeria drupacea
A much-branched, bushy, honey-scented shrub in coastal areas and light bush. Leaves reduced to minute scales, so that plant appears leafless with long, slender, upright green twigs. Flowers very small under 3 mm across, very numerous in spike-like inflorescences on the twiggy branches. Under a lens each flower is seen to have a perianth with 5 white segments and 5 short stamens arising between the 5 lobes of the nectar disc. So much nectar is produced that the honey smell is noticeable from a distance of several metres. Flowering September-December. Tas, NSW, Qld.
Santalaceae

Eucalyptus globulus *Blue Gum*
A large tree to 60 m under good conditions, with a skirt of rough bark, smooth and white on upper trunk and limbs. Leaves, young buds and fruits blue-green and with white bloom. Easily identified by its very large blue-green juvenile leaves on square stems, its large, up to 30 cm long, dark green, sickle-shaped adult leaves, smelling strongly of eucalyptus, and the large, blue-green 4-ribbed capsule 2 cm across, covered with white waxy bloom. Flower large, 15-20 mm across, solitary or 3 together in the leaf axils. Flowering spring-summer. Common in the south, south-east, east, Flinders and King islands. Tas, Vic (Wilsons Promontory).
Myrtaceae

Helichrysum purpurascens
Neat columnar shrub 1-2 m high, with dense aromatic foliage. Leaves are narrow-linear, blunt, margins recurved, smooth above, white beneath, 7-18 mm long. Flower heads in many dense clusters at ends of main and side branches. Buds often purplish-pink or brownish, the colour of the shortly hairy outer bracts of each flower head; flower heads white, inner bracts white tipped, 8-10 disc florets. Pappus bristles with minute barbs. Flowering December-January. Common on dry open hillsides in the south, sea level to about 250 m. Tas, endemic.
Asteraceae

181 *Eucalyptus pauciflora*
Cabbage Gum, Weeping Gum

182 *Eucalyptus pauciflora*
Cabbage Gum, Weeping Gum

183 *Rubus parvifolius* Native Raspberry

184 *Leptomeria drupacea*

185 *Eucalyptus globulus* Blue Gum

186 *Helichrysum purpurascens*

Pultenaea daphnoides var. **obcordata** *Native Daphne*
An erect shrub with semi-erect branches to 3 m high. Leaves cuneate, flat, 1-2 cm long, up to 1 cm wide, mid-rib produced into a small abrupt point, often light green, usually glabrous, underside paler. Flowers in terminal heads surrounded by small bracts, flowers golden yellow with reddish markings on the standard and purplish-brown keel almost hidden by the wings. Pod ovate, short, softly hairy. Flowering spring. Common, in shaded habitats and on hillsides at low altitudes. Tas, Vic, SA.
Fabaceae-Faboideae

Lomatia tinctoria *Guitar Plant*
Small stiff shrub 0.5-1 m or more, sparsely branched, spreading by rhizomes. Leaves pinnately divided once or twice into long slender leathery lobes. Flowers long stalked, white or cream, in long terminal racemes. Buds curved, open flower with 4 recurved perianth lobes ±10 mm, tips bearing sessile stamens. Ovary long stalked, style curved. Fruit dry, opening along one side into an oval boat or guitar-shape. Seeds winged. Flowering January. Widespread on hillsides and in light forest. Endemic.
Two other species of *Lomatia,* both endemic in Tasmania. *L. polymorpha* is illustrated in Montane section (plate 54). *L. tasmanica* is a small tree with pinnate leaves, leaflets lobed, prickly edged. Flowers crimson, petals fleshy. Rare, in rainforest of the far south-west.
Proteaceae

Astroloma pinifolium
Small stiff shrub, erect or spreading, 20-60 cm tall, one coastal form prostrate. Leaves 12-25 mm long, 1-2 mm wide, crowded, needle-like, with recurved margins, often pressed upwards and erect on the last two season's growth, lower parts of the stem almost bare. Flowers long yellow tubular bells, 15-18 mm long constricted at the throat, the free tips of the petals densely bearded on inner surface and curved back. Fruit a spherical drupe. Flowering September-March. In coastal heaths and inland in north, east and south. Tas, Vic, NSW.
Epacridaceae

Caladenia congesta *Black-tongued Caladenia*
Rather rare plant easily recognised by its large 3 cm diameter deep pink flowers, and by the narrow central lobe of labellum with its densely packed purple-black glands. Slender, slightly hairy glandular plant with erect flowering stem, 15-25 cm high and very narrow leaf, shorter than stem. Flowers 1-3, about 3 cm across, bright deep pink, narrow sepals and petals and 3-lobed labellum, side lobes curved, erect, central lobe completely covered with glands. Flowering late spring. Fairly rare in wet places in eucalypt woodlands and high rainfall areas in the north. Tas, Vic, NSW, SA.
Orchidaceae

Olearia stellulata
A variable woody shrub, closely related to *O. phlogopappa,* 1.5 m tall with slender upright branches. Young growth, branches and underside of leaves densely covered with felty stellate hairs. Leaves oblong to lanceolate 5-9 cm long, 7.5-20 mm wide, margins usually coarsely and evenly toothed, upper surface reticulate, rough to the touch, dark green, under surface fawn with dense covering of stellate hairs. Flower heads numerous on long stalks, 1 or 2 together terminating short lateral branches, florets white, occasionally blue. Fruit with pappus, achenes hairy. Flowering September-February. Widespread, in margins of wet forests especially in the south. Tas, Vic, NSW, Qld.
Asteraceae

Indigofera australis *Native Indigo*
A slender often leggy bush 0.5-1.5 m tall with clusters of small pink-lilac pea flowers, occasionally white. Leaves blue-green, pinnate with 4-10 pairs of narrow leaflets and terminal leaflet. The pod is brown, narrow, ±cylindrical, 1.5-4 cm long, with 5-10 seeds. Flowering November. In sandy heaths and on hillsides in areas of light or moderate rainfall. All States.
Fabaceae-Faboideae

187 *Pultenaea daphnoides* var. *obcordata*
Native Daphne

188 *Lomatia tinctoria*
Guitar Plant

189 *Astroloma pinifolium*

190 *Caladenia congesta*
Black-tongued Caladenia

191 *Olearia stellulata*

192 *Indigofera australis* Native Indigo

Exocarpos cupressiformis *Native Cherry*
A shapely small tree to 8 m with brown ridged bark and numerous close green or greenish bronze twigs. Leaves reduced to minute scales, persistent. Flowers 6-8 in small lateral and terminal axillary spikes 3-6 mm long, about 1 mm across, having a minute perianth with 5 white lobes. Usually only one flower of the spike develops into a fruit; the ripe ovary is perched on the thickened flower stalk which becomes succulent, red and edible, about 5 mm long. Species of *Exocarpos* are parasites on roots of other plants, so these attractive trees are difficult to transplant or grow from seed. Flowering in spring. Widespread in light forests. Tas, temperate. Australia.
Santalaceae

Acacia dealbata *Silver Wattle*
Tree to 30 m tall with grey mottled smooth bark and felty or hoary branchlets, flowering when very small. Leaves bipinnate, minutely hairy, blue-green, with 10-20 pairs of pinnae, raised glands at base of each pair; smallest leaflets in 30-40 pairs, linear, crowded, 4-6 cm long. Globular heads each of 25-30 flowers, lemon or bright yellow on hairy stalks 5-6 mm long, in long racemes or panicles. Pods light purplish-brown, oblong, smooth, 5-8 cm long, 8-10 mm broad, in good seasons forming conspicuous masses. Flowering July-August. Widespread from coast to mountains, largest in moist valleys and colonising after fires or other disturbances in eucalypt forests. Tas, Vic, NSW, introduced into SA.
Fabaceae-Faboideae

Viola hederacea *Wild Violet*
An attractive small perennial herb with long-stalked white flowers with a mauve centre, held well above the leaves. Flowers irregular like garden violets but not spurred and not perfumed. The plants spread by creeping stolons and may cover the ground with a mat of stalked kidney-shaped leaves, 5-15 mm long with slightly toothed margins. Flowering for many months in spring, summer and autumn. Abundant from coastal bush and lowland forests to alpine areas. Tas, Vic, NSW, Qld, SA; Malaysia.
Violaceae

Pimelea filiformis
A prostrate or scrambling plant forming small flat clumps, bearing clustered pale pink flowers on very slender stems. Bark light brown to grey, leaves narrow elliptical, light green, paler underneath, 6-12 mm long. Flowers in clusters of 6-9, each flower tubular 5 mm long with one stamen only. Flower tube and buds deep pink. Fruit dry. Flowering November-December. Very local in distribution in northern Tasmania, on hills each side of Tamar River. Tas, endemic.
Thymelaeaceae

Helichrysum apiculatum
A yellow-flowered everlasting daisy with branched inflorescences of many small flower heads. Plant rhizomatous with several stems arising from the woody base. Leaves grey-green with woolly white hairs, lanceolate. Flower heads 4-7 in terminal inflorescences, a few centimetres across. Heads golden yellow, bracts bluntly pointed, florets yellow, rather long. Pappus present. Flowering spring-summer. Widespread. Tas, Vic, NSW, SA.
Asteraceae

Bursaria spinosa *Prickly Box*
A white flowered erect shrub or tree to 10 m, upper part much branched, the smaller branches and axillary shoots ending in spines. Leaves oblanceolate or obovate, rather thin, blunt, 1-5 cm long. Numerous white flowers in branched pyramidal inflorescences; the sepals are shed early. The flower then consists of 5 narrow oblong petals, 5 stamens and a central ovary. Fruit a heart-shaped flattened, thin textured, brittle capsule. A source of nectar for honey production. Flowering November-January. Widespread, common especially on dry hillsides, rocky places and coastal sands. All States.
Pittosporaceae

193 *Exocarpos cupressiformis* (with fruits)
Native Cherry

194 *Acacia dealbata*
Silver Wattle

195 *Viola hederacea* Wild Violet

196 *Pimelea filiformis*

197 *Helichrysum apiculatum*

198 *Bursaria spinosa* Prickly Box

Convolvulus erubescens *Pink Bindweed*
A slender trailing or twining plant growing from persistent rootstock. Leaves variable, lower ones hastate with toothed or lobed margins, upper ones narrower, linear-oblong. Flowers solitary, pink or white, petals joined to form a wide funnel-shaped corolla, up to 20 mm diameter. Fruit a capsule. Flowering spring and summer. Common in dry open places, roadside banks. All States.
Convolvulaceae

Hypericum gramineum *Small St John's Wort*
Slender hairless yellow-flowered herb, with perennial base and opposite oblong-lanceolate leaves, stem clasping at the base. Flowers 1.5 cm across, solitary, terminal in a cymose inflorescence. Stamens numerous, more than 20, free and spreading, surrounding the central ovary. Fruit capsular. Flowering spring. Common in pastures and grassy places. Tas, most mainland States; New Zealand, New Caledonia.
Clusiaceae

Dodonaea filiformis *Fine-leaved Hop-bush*
Conspicuous when in fruit this small erect much-branched shrub, 1-2 m high, has male and female flowers on separate plants. Leaves fine narrow-linear, crowded, sticky, usually 1-2 cm long. Flowers in small terminal and axillary groups. Male flowers with thick oblong stamens, female with 3-lobed ovary. Numerous small papery reddish-brown fruits, usually 3-winged, broader than long. Flowering December. Usually found on well-drained slopes and on river banks. Tas, endemic.
Sapindaceae

Helichrysum scutellifolium
'Leaves like little shields' is the meaning of the specific name. The tiny dark green slightly sticky leaves are circular raised bumps, closely pressed to the stem, almost embedded in the white woolly hair which clothes the many slender branches. The flower heads 3-5 together, right at the end of the branches, appear dirty because of their outer brownish bracts. Florets are yellowish-white about 4 mm long, quite large in comparison with the leaves; pappus bristles straw coloured. Flowering spring. Local on hillsides in various parts of the State. Tas, endemic.
Asteraceae

Acacia axillaris
A bushy densely branched shrub 3-4 m tall. Phyllodes dull green, narrow, up to 5 cm long, 1-1.5 mm wide, angular and sharply pointed with a central vein. Stalkless heads of 3-4 tiny flowers, light to medium yellow, several together in the leaf axils; buds rounded. Pods small, brown, flat, 3-4 cm long, 2-3 mm wide. Flowering October. Known only from flats and valleys of the Elizabeth, St Pauls and Clyde rivers. Tas, endemic.
Fabaceae-Mimosoideae

Clematis gentianoides
An erect non-climbing Clematis, with many short branching stems to 45 cm high, arising close together from persistent rootstock. Leaves simple, opposite, lanceolate, about 8 cm long, veins pinnate but appearing longitudinal. Flowers terminal on the erect branches, long stalked, usually male and female on separate plants, occasionally hermaphrodite. Flowers white, to 6 cm across, with 4-8 petaloid sepals; no petals. Male flowers have numerous stamens with short points; female less showy with sepals and many carpels in a dense head, styles plumose, about 2.5 cm long in fruit. Flowering November-January. Scattered widely especially in the north and east on rocky hillsides. Tas, endemic.
Ranunculaceae

199 *Convolvulus erubescens*
Pink Bindweed

200 *Hypericum gramineum*
Small St John's Wort

201 *Dodonaea filiformis* (fruits)
Fine-leaved Hop-bush

202 *Helichrysum scutellifolium*

203 *Acacia axillaris*

204 *Clematis gentianoides*

Acacia mearnsii *Black Wattle*
Tree to 10 m tall with dark grey rough bark; branches widely spreading or drooping, sometimes touching the ground; branchlets angular, hairy. New growth golden bronze. Bipinnate leaves with 12-20 pairs of pinnae on angular stems with raised glands at base of each pair of pinnae. Smallest leaflets in 30-50 pairs, minutely hairy, crowded, linear, 1.5-4 mm long. Fragrant pale yellow globular heads of 20-30 flowers on golden-hairy stalks, 5-8 mm long, in long axillary racemes or panicles. Pods dark brown, flat, 5-10 cm long, 5-8 mm wide, slightly constricted between seeds. Bark used in tanning. Flowering November-December. Common and widely distributed. Tas, Vic, NSW, Qld, SA.
Fabaceae-Mimosoideae

Stenanthemum pimeleoides
Prostrate plant with woody base and slender branches 15-30 cm long, forming small mats. Leaves 3-6 mm long, are shining dark green, paler and slightly hairy underneath, obovate or obcordate with notched apex. Flowers white, about 2 mm wide, tightly packed in small round flat or domed terminal heads with brown bracts outside, surrounded by 2-3 conspicuous white softly hairy leafy bracts, the same shape as the foliage leaves but larger. Flowering November-December. Local in drier parts of midlands and east, often on stony ground or on ironstone gravel. Tas, endemic.
Rhamnaceae

Pimelea nivea *Round-leaf Rice Flower, Cotton Bush*
A shrub to 2 m with slender leafy branches terminated by white or pinkish heads of flowers. Leaves small, opposite, elliptical, often as broad as long, dark green, shining with dense white woolly hairs beneath and along young branches. Flowers tubular in dense heads of 10 or more on main and lateral branches, tubes silky hairy, lobes short, spreading, stamens 2. Fruits dry surrounded by long hairs. Sometimes called Bushman's Bootlace from the tough stringy bark which can be stripped from the branches. Flowering October-December. Common and widespread on rocky hillsides to 1000 m. Tas, endemic.
Thymelaeaceae

Olearia ramulosa *Twiggy Daisy-bush*
A slender shrub to 2 m with long straight shoots and a few lateral branches. Leaves are variable, narrow-linear to oblanceolate 4-8 mm long, flat, margins recurved, often crowded by development on small lateral branches; leaves sometimes very small but in one coastal form are to 15 mm long. Flowers small, white, to 1 cm across, sessile, axillary, numerous, with few ray florets, few disc florets. Fruit with pappus. Flowering spring-summer. Widespread behind dunes, in wet places in heaths and woodlands, and on hillsides in moderate rainfall areas. All States, not NT.
Asteraceae

Tetratheca pilosa *Lilac Bells, Black-eyed Susan*
Shrubby plant to 60 cm, widespread in heaths and dry forests. Mauve or purple flowers, 4 petals, black stamens, often crowded along the upper part of the stem forming showy heads. Sepals lost as flower opens. Leaves narrow-linear, about 10 mm long, glabrous or hairy, soft with recurved margins. Flowering spring-summer. Widespread and common. Tas, Vic, NSW, SA.
T. procumbens (not illustrated). Like *T. pilosa* but all parts very small, to 10 cm high. Wet forests and montane. Endemic.
Tremandraceae

Tetratheca labillardierei
A branched undershrub to 60 cm; leaves, stems, flower stalks and calyces glandular hairy. Flowers mauve, stalked, like those of *T. pilosa* but larger and with persistent sepals. Leaves narrowly ovate, slightly narrowed into very short stalk, glandular, margins slightly revolute or toothed. Flowering spring-summer. Widespread in heaths and light forests from sea level to mountain foothills. Tas, Vic.
T. ciliata (not illustrated). Flowers large 1 cm long, leaves ovate not glandular, in whorls around the stem. Very local near north coast.
Tremandraceae

205 *Acacia mearnsii* Black Wattle

206 *Stenanthemum pimeleoides*

207 *Pimelea nivea*
Round-leaf Rice Flower, Cotton Bush

209 *Tetratheca pilosa*
Lilac Bells, Black-eyed Susan

208 *Olearia ramulosa* Twiggy Daisy-bush

210 *Tetratheca labillardierei*

Burchardia umbellata *Milkmaids*

A white-flowered lily 15-50 cm high with flowers in umbels, and hairless grass-like leaves with large sheathing bases. Roots fibrous; several large leaves to 20 cm long at the base of the plant, 1 or 2 smaller on the stem. Umbel terminal, occasionally a second umbel in the axil of top stem bract. Flowers with 6 white 'petals', 6 conspicuous stamens and a purplish-pink ovary, triangular in section in centre of the flower. Named in honour of the German botanist J. H. Burckhard (1676-1738). Flowering spring. Common in grasslands, open forests and sandy coastal areas. Temperate Australia.
Liliaceae

Diuris longifolia *Wallflower Diuris*

Named Wallflower Diuris because of its flower colour, a soft blend of brown, purple and golden yellow, this plant has 2 or 3 long grass-like leaves and an erect stem to 45 cm. Flowers 1-6, usually 2 or 3 on slender stalks in terminal raceme. Flower rather broad, two upper clawed petals large, rounded and erect; central dorsal sepal wide; labellum with 2 spreading narrow side lobes and rounded, convex middle lobe, all about the same length; lower sepals long, strap shaped, sometimes crossing below the flower. Flowering September-November. Widespread, mainly in north and east. Tas, Vic, NSW, SA, WA.
Orchidaceae

Hovea linearis

Small shrubby pea with slender branches, drooping and trailing through other vegetation. Leaves elliptical to narrow-lanceolate, alternate, about 3 cm long with slightly recurved margins and prominent mid-veins. Flowers blue-mauve, 8 mm across, 1-3 in the leaf axils along most of stem in a leafy inflorescence. Calyx brown, unequal, 2 upper lobes broad, the 3 lower narrow. Pod 2-seeded, about 1 cm long, almost spherical. Flowering spring. Widespread in dry grassland and stony places. The genus is endemic in Australia. Tas, Vic, NSW, Qld, SA.
Fabaceae-Faboideae

Bulbine bulbosa *Bulbine Lily*

Perennial herb with bulbous rootstock and thickened roots. Leaves basal, grass-like, erect, rather fleshy, to 30 cm long. The unbranched erect flower stalk 8-60 cm long bears numerous flowers, 1.5-2 cm in diameter, in an elongating raceme with many unopened buds at the apex, and setting fruit at the bottom. Flower of 6 bright yellow petals, a cluster of 6 bearded stamens, and central ovary. Fruit a spherical capsule about 6 mm wide, with black angular seeds. Flowering October-December. Widespread in open rocky, sandy or grassy areas from sea level to mountains. Tas, Vic, NSW, Qld, SA.
Liliaceae

Billardiera scandens *Apple Dumplings*

The common name of this small woody plant with twining branches comes from the edible yellow fruit. Leaves linear-lanceolate, sometimes wide, 3-5 cm long, the margins often very undulate. Upper surface dull green, undersides paler with prominent mid-ribs. Pendulous bell-shaped flowers, greenish-yellow, becoming purplish with age, 16-25 mm long, the 5 petals forming a tube then spreading widely, the tips recurved. Flower stalks slender, 2.5 cm long. Fruit a slender cylindrical berry, 15 mm long, olive-green becoming yellow when ripe. Flowering October. Common in the north in dry woodland. Tas, Vic, NSW, SA.
Pittosporaceae

Solanum laciniatum *Kangaroo Apple*

Although this plant often appears to be a very large herb it may grow to a shrub of 3 m. Leaves variable in shape from lanceolate with a few coarse lobes near the base, to narrow-lanceolate 10-25 cm long. Long stalked large purple flowers 5 cm across, are borne in few flowered racemes, their petals joined in a wide flat spreading corolla, stamens prominent. Fruit a drooping ovoid berry 3 cm long, orange-yellow when ripe. Flowering for several months in summer. Common in damp shaded places, often colonising disturbed patches. Tas, Vic, NSW, SA; New Zealand.
Solanaceae

211 *Burchardia umbellata* Milkmaids

212 *Diuris longifolia* Wallflower Diuris

213 *Hovea linearis*

214 *Bulbine bulbosa* Bulbine Lily

215 *Billardiera scandens* Apple Dumplings

216 *Solanum laciniatum* Kangaroo Apple

Pomaderris elliptica *Yellow Dogwood*

Shrub or small tree often flowering when very small. Leaves elliptic-lanceolate, leathery, 1-4 cm long, veins impressed on upper surface, dark green, fawn underneath with stellate hairs. Flowers small varying from lemon to golden yellow in large heads at the ends of the branches. Individual flowers have 5 hairy sepals, 5 thin yellow upstanding petals, 5 stamens longer than petals, style 3-lobed protruding from domed nectar disc. Flowering September-November. Common on rocky hillsides in north and north-east. Tas, endemic.
Rhamnaceae

Clematis aristata *Clematis*

Woody climber scrambling over shrubs, fallen logs and fences. Leaves opposite, trifoliolate, sometimes variegated, long stalked, each leaflet lanceolate, sometimes toothed, to 8 cm long. Flowers creamy white or pinkish, starry with 4-7 long narrow petaloid sepals. Male and female flowers on separate plants. Male flowers more showy with numerous pointed stamens. Female with many carpels with feathery styles persisting in the fruit. Flowering spring. Widespread in light forest and gullies. Tas, Vic, NSW, Qld, WA.
Ranunculaceae

Epacris virgata

A small white heath up to 60 cm high with several slender upright branches. Leaves narrow-oblanceolate, spreading or semi-erect 4-6 mm long, flat, tips pointed but not prickly. Delicate white tubular flowers, solitary in numerous axils along long lengths of stem or terminating short branches. Bracts and sepals ovate, corolla tube as long as calyx, lobes spreading with blunt tips, flowers 6 mm across, anthers protruding, dark coloured. Flowering September-October. Restricted to damp places in foothills of the Asbestos Range north of Launceston. Tas, endemic.
Epacridaceae

Spyridium obcordatum

A small prostrate shrub with wiry branches forming dense mounds and mats. Flowers are small, 3-6 together in dense heads surrounded by brown bracts and silky hairs; the small shining dark green leaves are about 5 mm long and almost as broad with a deeply notched apex, white, silky hairy beneath except for the prominent mid rib. Flowering July-August. Locally abundant in several areas near Beaconsfield in northern Tasmania, and near Spring Bay. Tas, endemic.
Rhamnaceae

Wurmbea dioica *Early Nancy*

A small bulbous rooted plant with simple stems 10-20 cm high, bearing 2-6 white flowers opening very early in spring. Leaves few, to 10 cm long, hairless, leaf base stem clasping, blades spreading, tapering to a long point. One to 8 flowers sitting closely on main stem; 6 perianth members (3 sepals, 3 petals) white, each with a band of purple towards the base and together forming a dark ring. Male and female flowers on different plants; male with 6 stamens, female with a 3-lobed purple-black ovary. Flowering July-September. Common in pastures, grasslands and light forest. All States and Central Australia.
A second species *W. uniflora* has only one flower and flowers later in spring.
Liliaceae

Wahlenbergia sp. *Bluebell*

There are several species of *Wahlenbergia* varying in flower size and length of corolla tube. Annual or perennial herbs about 50 cm high, usually branched near the base with slender stems. Leaves confined to the lower part of stem. Flowers bright blue with 4, 5 or more petals, bases joined into short tube and lobes spreading. Calyx of narrow ± spreading lobes which in the species illustrated are slightly longer than the corolla tube. Fruit a capsule. Flowering spring-summer. All States.
Campanulaceae

217 *Pomaderris elliptica* Yellow Dogwood

218 *Clematis aristata* Clematis

219 *Epacris virgata*

220 *Spyridium obcordatum*

221 *Wurmbea dioica* Early Nancy

222 *Wahlenbergia* sp. Bluebell

Acianthus reniformis *Mosquito Orchid*

Small ground orchid, stem erect 3-10 cm high, one orbicular pale green leaf 1-3 cm across closely pressed to soil. One to 5 small pinkish brown or greenish flowers. Labellum narrow strap shaped, about 12 mm long, with 2 faint shining longitudinal bands reminiscent of a mosquito with wings at rest. Dorsal sepal arched or erect, shorter than labellum, lateral sepals and petals shorter and narrower. Column about 6 mm long, round at the end and bent like a comma. Flowering July-September. Common on hillsides, in forests and dry heathy areas. All States.
Orchidaceae

Brunonia australis *Blue Pincushion*

Named for the early botanist Robert Brown, this beautiful blue-flowered perennial herb once plentiful in the north is now restricted to a few widely separated localities. Flower heads 1 or more on separate stalks arise from a rosette of greyish, slightly spathulate leaves; both leaves and stalks are softly hairy. The small individual florets of the head have 5 deep sky blue petals, paler on the back. Anthers form a tube around the purple style which protrudes above the level of the petals, making the pins on the cushion. Flowering November-December. Sandy or gravelly soil in dry forests in the north. All States.
Goodeniaceae

Pimelea humilis *Common Rice Flower*

A small plant to 30 cm high, tufted with many erect branches. Leaves about 1 cm long, sessile, dark green, in opposite pairs, sometimes forming 4 rows. Flowers white, tubular in erect heads at the ends of the branches and surrounded by 4-6 leaf-like bracts. The outer surfaces of the flowers are covered with very short hairs. Short hairs at base of flowers persist around the pointed dry fruits. Flowering spring-summer. Widespread in sclerophyll forests and fairly dry places. Tas, Vic, NSW, SA.
Thymelaeaceae

Dichopogon strictus *Chocolate Lily*

A mauve-flowered chocolate scented lily with tuberous roots and basal grass-like leaves to 25 cm long. Stem erect, occasionally branched forming two flowering racemes. Flowers long stalked, the perianth of 6 separate segments, the inner 3 with fringed margins and broader than the outer 3; flowers in elongated terminal racemes. Stamens 6 with 2 very short hairy lobes at base of each oblong anther. Anthers purplish-black and conspicuous, ovary rounded. Capsule erect 7-8 mm long. Flowering November-December. Widespread in grassland or woodland. Tas, Vic, NSW, Qld, SA.
Liliaceae

Chrysanthemoides monilifera *Bone Seed*

A shrub to 3 m with dark green toothed or lobed fleshy leaves covered with white hairs which remain as webby fragments on older leaves. Terminal heads of yellow daisies with only 5-7 ray florets and large yellow discs, producing clustered ovoid black fruits without pappus, but succulent and spread by birds. Originally introduced as a garden plant and soil binder, it has become an invasive problem weed capable of replacing native vegetation in sandy and rocky soil, by efficiently absorbing all available moisture. Strong eradication measures are necessary. Flowering spring-summer. Native of South Africa.
Asteraceae

Pterostylis plumosa *Bearded Greenhood*

This greenhood is easily recognised by its long hairy yellow tongue, the labellum. Leaves numerous, lanceolate, pointed, in a rosette at base of stem and often on stem itself. Flowers solitary, erect, large, 2.5 cm long; stem up to 30 cm long. The boat-shaped hood is erect, inflated at the base and pinched in near the middle, shortly pointed at the apex. The two narrow, yellow-orange lower sepals about as long as the hood are joined at the base then turn down and out; column is hidden in the hood; thread-like labellum about 2 cm long with many fine long yellow hairs and small terminal knob, touch sensitive. Fruit, large erect capsule with many very small seeds. Flowering September-November. Widespread in light forest or heath country. Tas, Vic, NSW, SA; New Zealand.
Orchidaceae

223 *Acianthus reniformis* Mosquito Orchid

224 *Brunonia australis* Blue Pincushion

225 *Pimelea humilis* Common Rice Flower

226 *Dichopogon strictus* Chocolate Lily

227 *Chrysanthemoides monilifera* Bone Seed

228 *Pterostylis plumosa* Bearded Greenhood

Eriostemon verrucosus *Wax-flower*

A small shrub to 80 cm, sprawling or erect. Leaves obovate or obcordate, concave, sometimes folded about mid-rib, warty with large oil glands, aromatic, upper surface less warty. Flowers numerous, 1.5 cm across, solitary in leaf axils near ends of branches, white or pink, with five or more petals, stamens straight, clustered in ring in the centre of the flower. Fruit dry. A form with much larger 1.5 cm folded leaves occurs near Mt Amos, Freycinet National Park. Flowering spring-summer. Common on dry rocky hillsides. Tas, Vic, NSW, SA.
Rutaceae

Daviesia latifolia *Native Hop, Bitter Leaf*

A common pea flower with very numerous small brown and yellow flowers in axillary racemes and broad leathery leaves. Shrub 1-2 m high, branches angular, spreading. Leaves broadly elliptical or obovate, dark olive green, leathery, network of veins visible on both surfaces, margins crenate. Racemes of flowers 3-6 cm long, 2-3 cm across, each flower small, purplish-brown and yellow about 4 mm across, subtended by a small bract. Pod triangular, flat, to 10 mm. The name Native Hop may have arisen from papery appearance of masses of flat young developing pods and the bracts beneath them. Leaves and heads of flowers were used to make infusions in the early days of settlement, used as medicine, or a bitter tasting substitute for hops, reported to have a purgative effect. Flowering October-November. Widespread and common on roadside banks. Tas, Vic, NSW, Qld.
Fabaceae-Faboideae

Sarcochilus australis *Gunn's Tree Orchid, Butterfly Orchid*

Tasmania's only epiphytic orchid is a very small plant less than 10 cm across. Leaves narrow, up to 7.5 cm long, pale green, growing on branches and trunks of trees. Several up to 10 small flowers, greenish-white lightly marked with purple-red, are carried on a drooping stalk. Although plants are small, the aerial roots can extend for up to 1 m along branches. Named for Ronald Campbell Gunn, a noted Tasmanian plant collector in the mid 1800s. Flowering summer. Grows in wet forest areas, and fern gullies. Not common but widespread. Tas, Vic, NSW, Qld.
Orchidaceae

Caladenia menziesii *Hare's Ears*

Slender ground orchid with erect flowering stem 5-20 cm high. Leaf pale green, broad-oval to oblong-linear, up to 2.5 cm long, without hairs. Pink and white flowers, 1-3 together with deeper pink or crimson markings. The 2 clubbed erect dark red petals, much longer than the lower sepals, form the ears. Crimson hood over column, its outer surface glandular hairy; lateral sepals white or pale pink with central crimson streak underneath; blade of labellum pointing forward, sides barred with red streaks, calli in 2-4 rows. Flowering October-November. Widespread, local in light coastal soils and light forest, especially after fires. Tas, Vic, SA, WA.
Orchidaceae

Ptilotus spathulatus *Pussy Tails*

A perennial herb with spreading prostrate stems ending in fluffy flower heads. Spathulate leaves in a basal rosette with shorter lanceolate leaves on the stems. The fluffy heads are erect spikes of sessile flowers each with a shining papery bract and 5-lobed perianth about 12 mm long, its outer surface densely covered with long straight hairs, the inner surface hairless. Flowering September-January. Occasional in dry grassy places. Temperate Australia.
Amaranthaceae

Linum marginale *Wild Flax*

This slender herb 30-60 cm tall has blue flowers, smooth slender blue-green hairless stems and narrow leaves 5-25 mm long. Flowers 5-petalled, blue, long stalked in terminal clusters. Stamens 5, joined at the base, ovary of 5 fused carpels with spreading stigmas. Flowering spring-summer. Widespread and common. Temperate Australia.
Linaceae

229 *Eriostemon verrucosus* Wax-flower

230 *Daviesia latifolia* Native Hop, Bitter Leaf

231 *Sarcochilus australis*
Gunn's Tree Orchid, Butterfly Orchid

232 *Caladenia menziesii*
Hare's Ears

233 *Ptilotus spathulatus* Pussy Tails

234 *Linum marginale* Wild Flax

Caladenia patersonii *Common Spider Orchid*
Flowers of this spider orchid are large, 5 cm across; colour varies from pale cream through shades of yellow and brown to deep red. One narrow hairy pointed leaf, with a hairy stem bearing 1, 2 or 3 flowers. Sepals and petals spreading, 5-8 cm long, the tails darker with numerous glandular hairs. Labellum margins fringed with coarse teeth progressively shorter towards the tip which is curved under; 4-6 rows of calli on labellum. Flowering August-November. Widespread and often abundant in lightly forested areas. All States.
Orchidaceae

Dipodium punctatum *Hyacinth Orchid*
Leafless erect orchid 30-90 cm tall, saprophytic, leaves replaced by brown scales so that the shoots at first look like brown asparagus. Stem fleshy but hard, usually dark purplish-brown, rarely green. Flowers pink about 3 cm across with deeper pink spots, occasionally unspotted, on slender short stalks, petals spreading and recurved, bluntly pointed. Flowering summer. Widespread in open forest areas especially near the coasts. Found in close association with trees, usually eucalypts, but is not parasitic. Tas, Vic, NSW, SA.
Orchidaceae

Thelymitra aristata *Great Sun Orchid*
In Tasmania this plant was formerly known as *T. grandiflora.* A robust orchid with many flowered erect stem to 80 cm high, sometimes smaller and slender, one thick ribbed sheathing leaf at base and several small bracts on stem. Flowers few or numerous in a long raceme, opening freely. Segments pale blue to pinky-mauve, often pink-brown on back of petals. Side lobes of column with terminal tufts of white or pale yellow hairs. Flowering late spring. Widespread in open eucalypt forests and peaty heaths. Tas, Vic, SA.
Orchidaceae

Acianthus caudatus *Mayfly Orchid*
Small ground orchid, one heart-shaped leaf, 1-2.5 cm broad, flat on the ground, upper surface dark green, reddish-purple underneath, erect stalk bearing 3-4 dark purplish-red flowers with very long filiform sepals, having an unpleasant doggy smell. Flowering early spring. Widespread in shady places in light forest, often coastal. Tas, Vic, NSW, SA.
Orchidaceae

Gastrodia sesamoides *Potato Orchid*
Leafless orchid growing from large underground tuber which may be 15 cm long x 8 cm wide. Stem robust, brown, to 120 cm high, usually about 80 cm. Flowers numerous in terminal raceme, individual stalks elongating as flower and fruit matures. Flowers drooping, tubular bell-shaped; tube split above, 5 lobed, brown outside, creamy white inside. It is reported that Tasmanian Aborigines roasted and ate the tubers of this plant. The plant lives on rotting humus in the soil with the help of fungi in its roots. Flowering October-December. Found in light forest in isolated clumps, very plentiful in some years. All States; New Zealand.
Orchidaceae

Chiloglottis gunnii *Common Bird Orchid*
Small tuberous plant with short erect flowering stems 4-8 cm high bearing 1 large terminal flower. Two spreading basal leaves, 2-7 cm long, ovate or broadly lanceolate with short stalks. Flower 2 cm wide, green to reddish-brown, opening widely like a bird's mouth. Upper sepal forming a short pointed hood, other sepals narrow, pointing forward below. Labellum cordate, sides higher than centre, reddish-brown, a large stalked black callus at base, a shorter thick one near centre, and smaller ones in two irregular rows on either side. Flowering in spring near coasts, to January in mountain areas. Widespread, usually in shade in forest areas, sea level to about 950 m. Tas, Vic, NSW, Qld.
Orchidaceae

235 *Caladenia patersonii*
Common Spider Orchid

236 *Dipodium punctatum*
Hyacinth Orchid

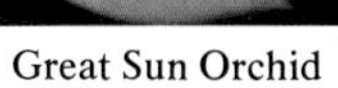

237 *Thelymitra aristata* Great Sun Orchid

238 *Acianthus caudatus* Mayfly Orchid

239 *Gastrodia sesamoides* Potato Orchid

240 *Chiloglottis gunnii* Common Bird Orchid

Eucalyptus perriniana *Spinning Gum*
Slender tree of mallee habit where fire is frequent, but reaching 7 m in favourable conditions. Bark, juvenile and adult leaves, buds and fruit glaucous. Juvenile leaves opposite, joined around the stem forming a disc which breaks free from the stem when old but remains on it, spinning in the wind, hence Spinning Gum. Adult leaves stalked, lanceolate or falcate, with slender points, about 9 cm x 1.5 cm. Flowers 3 together, axillary. Capsule deeply cup shaped, ± parallel sided, 5-6 mm across with raised rim. Often grown as an ornamental tree or for its foliage. Flowering summer. Local near Tunnack and Strickland at 300-600 m. Tas, Vic, NSW.
Myrtaceae

Eucalyptus morrisbyi *Morrisby's Gum*
An area at Risdon is reserved for the protection of this endemic tree which occurs naturally in only two small areas near Hobart. Its small size to 12 m makes it suitable for street planting and it is increasingly used as an ornamental. The bark is smooth greyish-white. The juvenile leaves are opposite, round, sessile, blue green with white waxy bloom (glaucous); adult leaves stalked, lanceolate 5-10 cm long, and glaucous. Flowers creamy-white; buds and fruits three together in leaf axils. Mature fruits cylindrical, about 1 cm long and nearly as wide. Very local at Risdon and near South Arm on sandy mudstone soil. Tas, endemic.
Myrtaceae

Eucalyptus risdonii *Risdon Peppermint*
Small open widely branched tree to 15 m tall, foliage silvery white, leaves covered with waxy bloom. Fresh bark on trunk and branches mottled in patches of white, silver and grey, old bark grey. Leaves, both juvenile and adult, blue-green, ovate-lanceolate with long points, opposite bases completely joined around the stem; overall length 6-10 cm. Small stalked cream flowers 5-15 together, in umbels, often several umbels in each leaf axil. Buds 5-8 mm long, glaucous, hemispherical with conical caps. Capsule 8-10 mm diameter, pear shaped, shining, smooth, flat across the top. Flowering irregularly. Tas, endemic.
Myrtaceae

Goodenia lanata *Native Primrose*
A small herb with a rosette of stalked toothed obovate leaves and long trailing sometimes rooting leafy stems. Flowers arising from the base of the plant and from the axils of leaves, on long stalks, 2 small bracts near middle. Flower yellow, 2 cm long, with small calyx, 5 yellow petals, each thick in the mid-line and thinner at the edge. Flower 2 lipped, upper 2 petals erect, lower 3 spreading, upper side of flower tube split almost to base. Pollen is shed into a cup at end of the style and remains until the stigma is receptive. Flowering spring-summer. Widespread in open dry forests. Tas, Vic, NSW.
Goodeniaceae

241 *Eucalyptus perriniana* Spinning Gum

242 *Eucalyptus morrisbyi* Morrisby's Gum

243 *Eucalyptus risdonii* Risdon Peppermint

244 *Goodenia lanata* Native Primrose

COASTAL HEATH

This section contains plants normally growing in low lying coastal areas on deep sandy soils low in nutrients, subject to exposure to winds and, in some cases, salt spray.

Correa alba *White Correa*
Unlike other Tasmanian species of *Correa,* the white petals of this shrub spread widely forming a cup-shaped rather than tubular bell flower. The leaves are broad, ovate to almost orbicular, 1-3 cm long, grey-green and leathery. Flowers 2-3 together on short stalks at the ends of branches, not pendulous, have 4 free thick petals, their outer surface with stellate hairs and 4 thick stamens with red anthers about as long as the petals. Fruit divided into 4 parts, shortly hairy. Flowering most of the year. Common in sandy soil on coasts, growing sometimes in sand. Tas, Vic, NSW, SA.
Rutaceae

Correa backhousiana
Bushy shrub 1.5 m high, young branches and undersides of leaves covered with rusty hairs. Leaves ovate or oblong, apex blunt or indented, upper surface glossy, pale rusty beneath, 1.5-3 cm long. Flowers wide, 1-3 together at ends of lateral branches, calyx small, cup like, edges recurved; flower tube pale whitish green, whole covered with yellowish stellate hairs which are denser and darker brown on the spreading lobes, 8 stamens, 4 with wide filament bases alternating with 4 narrow ones; ovary covered with short dense hairs. Fruit separating into four 2-seeded parts. Flowering spring-summer. Coastal, west of Rocky Cape, on west and south coasts and islands of Bass Strait. Tas, Vic.
Rutaceae

Alyxia buxifolia *Sea-box*
A sprawling coastal shrub often prostrate among rocks in the salt spray zone, but in more protected areas a 2 m high shrub with many spreading branches. The smooth shining leaves are oval 1-3.5 cm long, tough with a thick waxy skin. Scented white flowers in terminal clusters of 2-7 have a long corolla tube which is orange inside, with a little tight white rim at the throat, and white spreading lobes overlapping in rotation. Fruit an orange to red fleshy drupe 6-8 mm across. Flowering September-October. Local on rocks and cliffs in the north of the State. Tas coasts of temperate Australia.
Apocynaceae

Carpobrotus rossii *Native Pig Face*
Robust fleshy perennial plant with prostrate stems up to 1 m long. Leaves triangular in section, opposite, bases stem-clasping, tips pointed. These act as water storage organs enabling the plant to survive hot dry summers on coastal cliffs and sand. Flowers light purple 4-6 cm diameter, solitary on ends of short lateral branches, with a fleshy calyx and many shining narrow strap-shaped petals about 2 cm long in several rows. Stamens many, filaments white. Fruit fleshy, fig-like, yellowish when ripe, seeds small. Edible. Flowering spring and summer. Widespread, common in coastal areas on rocky headlands and sand dunes. Tas, Vic, SA.
Aizoaceae

Disphyma crassifolia *Round-leaved Pig Face*
Fleshy perennial herb spreading by prostrate stems and rooting at nodes so forming mats in salty soil at margins of salt marshes, and within spray zone on rocks and cliffs. Leaves rather club shaped, 2-5 cm long, rounded, thick and succulent with water storage tissue, blunt and often reddish at the apex. Flowers 3 cm across, terminal on short shoots. Petals very numerous, strap shaped, light magenta, white towards the centre, occasionally all white. Stamens numerous, white. Fruit succulent at first, becoming dry and opening by slits. Flowering spring and summer. Tas, temperate mainland.
Aizoaceae

245 *Correa alba* White Correa

246 *Correa backhousiana*

247 *Alyxia buxifolia* Sea-box

248 *Carpobrotus rossii* Native Pig Face

249 *Disphyma crassifolia* Round-leaved Pig Face

Senecio elegans *Purple Ragwort*

A purple-flowered coastal annual or short lived perennial herb. Leaves deeply bipinnately lobed, rather fleshy with base stem-clasping. Branching stems to 60 cm high, bearing terminal clusters of daisy flowers 2.5-4 cm across with bright purple rays and yellow discs. Fruit with pappus. A plant of stabilised sandy shores and low coastal banks especially on Bass Strait islands but not common on the mainland of Tasmania. Flowering October-December. Native of South Africa, has now spread in isolated patches along the coasts of WA, SA, Vic, Tas, and New Zealand.
Asteraceae

Calocephalus brownii *Cushion Bush, Snow Bush*

A rigid spreading shrub, with many wiry interlacing branches forming silvery mounds on coastal cliffs and headlands. Leaves narrow-linear 2-5 mm long, pressed to the branches, both leaves and branches white with blanketing hairs. Clusters of flower heads terminal, spherical, consisting of numerous tiny white heads of only 2 or 3 florets. Pappus bristles feathery. Flowering September-February. Locally common particularly in the north, forming wind pruned cushions to 1 m high on coastal dunes and cliffs. Coasts of temperate Australia.
Asteraceae

Acacia sophorae *Coast Wattle, Boobyalla*

A common shrub of the coastal sand dunes with long arching branches which root and bind the sand. Also found a short distance inland as a small tree up to 5 m tall. Phyllodes dark green, elliptical, blunt pointed with several prominent longitudinal veins. Flowers in long straight yellow spikes, 1 or 2 spikes together in the upper leaf axils. Pods thick, narrow, 8-15 cm long slightly narrowed between seeds. Flowering August-September. Coastal, sandy soil, sand dunes and fringing beaches. Tas, Vic, NSW, Qld, SA.
Fabaceae-Mimosoideae

Helichrysum reticulatum

A sturdy shrub with narrow-linear stalked leaves, dark green and wrinkled above, white underneath. Branches with dense white tomentum (fur). Flower heads white, nearly globose, 5-7 mm across, in flat or domed heads 5-7 cm in diameter. Buds and outer bracts of flower covered with white tomentum. Pappus with flattened minutely hairy tips. Flowering January. Coastal cliffs, rocky parts of sandy coasts, less common in north. Tas, endemic.
Asteraceae

Dodonaea viscosa *Native Hop Bush*

A tall shrub or small tree, 2-6 m high. Leaves oblong or spathulate, sticky, flat, green or reddish, broadest near the usually blunt apex. Terminal clusters of small flowers with 4 sepals, and 8 stamens, with thick oblong anthers, longer than the sepals. Fruit papery, reddish-brown, 3 winged, broader than long. Flowering November-December. Tasmania, widespread and abundant. All States; New Zealand; cosmopolitan in warm regions. A purple-leaved form is often grown for its foliage.
Sapindaceae

Lasiopetalum baueri *Slender Velvet Bush*

Small coastal shrub of very local distribution in three widely separated areas in the north. Leaves narrow-elliptical, blunt, stalked, pale grey-green with a dense covering of very short stellate hairs except on upper surface of the older leaves, upper surface has mid-rib impressed. Flowers stalked; stems, stalks and flowers covered with brown, yellow or white stellate hairs. Sepals 5, ± petalloid, pink or white, petals minute, stamens oblong, red, massed together in centre of the flower forming a red waxy five-armed star. Flowering spring. Tas, Vic, NSW, SA.
Sterculiaceae

250 *Senecio elegans*
Purple Ragwort

251 *Calocephalus brownii*
Cushion Bush, Snow Bush

252 *Acacia sophorae* Coast Wattle, Boobyalla

253 *Helichrysum reticulatum*

254 *Dodonaea viscosa* (flowers)
Native Hop Bush

255 *Lasiopetalum baueri*
Slender Velvet Bush

Calytrix tetragona
Erect branched shrub with many twiggy branches and small fine linear or terete leaves, glabrous or shortly hairy. Flowers solitary in axils near ends of upper branches, forming showy heads. Calyx, shallow, cup shaped, each sepal produced into a long fine awn projecting beyond the flower. Petals 5, white or pink. Stamens about 20, long, conspicuous. Fruit dry. Flowering October-December. Widespread in coastal heaths, where it is often dense and wind pruned, and along some northern rivers. All States.
Myrtaceae

Leucopogon parviflorus *Currant Bush*
Tall coastal shrub with semi-erect or spreading branches and grey bark. Leaves oblanceolate, 1.5-2.5 cm long, 4-6 mm broad, widest just above the middle, apex bluntly pointed, dull green, flat, margins slightly recurved with faint parallel veins. Flowers sessile, in numerous straight close spikes to 3 cm long, one spike per axil near ends of branches. Buds pinkish, flowers white with short tube, lobes spreading, densely covered with white hairs. Anthers at throat of tube. Fruit, cream waxy drupe. Edible, often eaten by seagulls. Flowering spring. Tas, temperate mainland.
Epacridaceae

Banksia serrata *Saw Leaf Banksia*
A small tree with stout grey trunk and branches; leaves large 8-15 cm long, leathery with serrate margins. Leaves spathulate or oblanceolate, veins conspicuous. Flower spikes large up to 20 cm long, 10 cm wide, flowers pale yellow to golden. Ripe capsules hard and woody, appearing as furry bulges on the sides of old cones. Seeds large, winged. Very local at Sisters Hills. Flowering spring-summer. Tas, Vic, NSW.
Proteaceae

Cyathodes abietina *West Coast Pink Berry*
An erect much-branched shrub to 2 m high with crowded leaves and large pink berries. Leaves large, 10-15 mm long, linear-lanceolate, tough and thick, grooved underneath and hard pointed, with some resemblance to foliage of *Abies* spp. (Spanish Fir). Flowers solitary, clustered near the ends of branches, white, tubular, both lobes and inside of tube hairy, stamens protruding. Fruit, pink fleshy drupe ± spherical about 2 mm diameter. Flowering October-December. Exposed rocky coasts in south-west and west, and on southern islands. Tas, endemic.
Epacridaceae

Kunzea ambigua
A white-flowered shrub with long arching branches bearing many sprays of flowers. The leaves are narrow-linear, dull green, often tufted on short lateral shoots. Flowers solitary in many axils on the main stem or terminal on the short lateral branches crowded into bottle-brush-like heads, honey scented. Each flower white, with 5 small sepals and petals and many long white stamens. Fruit small leathery capsules, shed before the next flowering season. Flowering October-December. Coastal heaths and wet scrub, in the granitic country of the east and north-east and on Bass Strait islands. Tas, Vic, NSW.
Myrtaceae

Selliera radicans
The fleshy spathulate bright green leaves 1-10 cm long of this small herb arise in clusters of 3-4 from a yellowish stem that runs along or sometimes under the ground. The flowers are solitary on stalks shorter than their accompanying leaf. The flowers are fan shaped, with 5 petals almost white on the inner surface but dull crimson or purplish-grey on the outer. In exposed positions the plant is only 2 cm high but in thick vegetation the leaves may be 15 cm long. Flowering October-November. Found in salt marshes, edges of tidal streams. Tas, Vic, SA; New Zealand and Chile.
Goodeniaceae

256 *Calytrix tetragona*

257 *Leucopogon parviflorus* Currant Bush

258 *Banksia serrata*
Saw Leaf Banksia

259 *Cyathodes abietina* (fruits)
West Coast Pink Berry

260 *Kunzea ambigua*

261 *Selliera radicans*

Westringia rigida
This handsome member of the mint family is a substantial bush to 3 m, flowering for a long period. Narrow-linear pointed leaves, 10-16 mm long, are glossy green above, white beneath except for the prominent mid-rib and revolute margins. Leaves in well-spaced whorls of 3, are more crowded near the ends of the branches, sometimes in whorls of 4. Mauve-white flowers 22 mm vertically, 16 mm wide, with magenta dots concentrated around the throat and on lower 3 petals, are borne in leaf axils towards the ends of branches. Fruit, 4 nutlets surrounded by the persistent papery calyx. Flowering spring-summer. Locally abundant on foredunes and rocky outcrops on the east coast, from Friendly Beaches to Southport Lagoon.
This plant bears scant resemblance to the mainland form of *W. rigida,* possibly being closer to *W. fruticosa* which occurs in similar habitats in NSW. Discussions as to its true identity are continuing.
Lamiaceae

Lepidosperma concavum
A common sedge with branched creeping rhizome and erect fans of linear leaves, flat on one side and convex on the other, 20-60 cm long. Flowering stalk erect, plano-convex like the leaves, but longer, ± 5 mm wide, margin slightly rough. Bract below flowers about as long as head, rather wide, ±folded, head dense with several narrow clusters of greyish brown spikelets, their bracts blunt with short points. Fruit about 1.5 mm in diameter, pale coloured. Widespread and abundant in coastal heaths and light forest. Tas, Vic, NSW, Qld.
Cyperaceae

Pelargonium australe *Wild Geranium*
A small plant with a persistent rootstock which forms erect leafy stems. Some coastal plants very robust. Leaves long stalked, reniform or ovate-cordate, margins crenate, growing from top of rootstock. Flowers several together in a stalked inflorescence, pinkish-white with reddish markings, 5 sepals, lower one slightly pouched, 5 free petals, the two upper larger than the others, ± 10 stamens, not all fertile, joined at the base surrounding a long 5-chambered ovary. Flowering spring. Widespread, in crevices between rocks, on rocky river banks and cliffs, coastal and inland to mountains. Tas, temperate mainland.
Geraniaceae

Thelionema caespitosum *Blue Grass Lily*
A lily with grass like leaves, to 30 cm high often in small tussocks. Flowers pale yellow or blue, star-like in a branched inflorescence, 6 perianth members, 6 stamens, yellow and bent in the open flower, stamens and anthers yellow, surrounding ovary. Fruit a capsule. Very common in coastal heaths and wet soaks. Flowering November-December. Tas, Vic, NSW, Qld, WA.
Liliaceae

Sowerbaea juncea *Vanilla Lily*
A small grass-like lily with a tuft of blue-green solid leaves, found only in the far north-east of this State but common in all mainland coastal heathlands as far north as Noosa in Queensland. Stems about 30 cm high bear a head of pale pinkish-mauve flowers, each with pale pink papery bracts and yellow stamens. The papery flowers retain their form for many months after picking. Seeds 3-6 in a pale brown 2-3 mm long capsule. Flowering spring. Near Gladstone, north-east Tasmania. Tas, Vic, NSW, Qld.
Liliaceae

Dillwynia glaberrima
Shrub to 1 m with slender arching branches. Leaves dark green usually glabrous, very narrow-linear almost terete but grooved on the upper surface and very shortly pointed. Standard bright yellow twice as wide as long with red markings. Keel and lateral petals yellow; flowers in short racemes in the upper axils or terminal making short heads. Pod short, rounded. Flowering spring-summer. Wet heaths both coastal and inland. Tas, Vic, NSW, SA.
Fabaceae-Faboideae

262 *Westringia rigida*

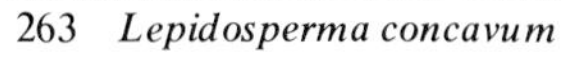

263 *Lepidosperma concavum*

264 *Pelargonium australe* Wild Geranium

265 *Thelionema caespitosum* Blue Grass Lily

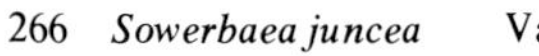

266 *Sowerbaea juncea* Vanilla Lily

267 *Dillwynia glaberrima*

Hakea teretifolia *Dagger Hakea*
A stiff much-branched shrub to 3 m high and wide. Leaves cylindrical, sharp pointed, stiff and straight 2-5 cm long. Flowers spider-like, creamy white, pubescent, clustered in groups in leaf axils. Fruit woody, elongated and dagger-shaped, broad with short lateral projections near the base then tapering to a long point about 3 cm long, opening to free 2 winged seeds. Flowering spring-summer. Common in coastal heaths. Tas, Vic, NSW, Qld.
Proteaceae

Paracaleana minor *Small Duck Orchid*
A tiny red or sometimes greenish orchid, often overlooked. Stem to 15 cm and wiry, leaf to 9 cm, narrow-linear, stem-clasping, often withered by flowering time. Flowers small usually 2-3, but occasionally 7. Five parts of the perianth are narrow, the column and its wings, reddish in colour, make the 'body' of the duck. Labellum at top of flower is covered with dark red raised glands and attached by a narrow flexible strap which acts as a hinge enabling the 'head' to swing down when an insect alights on the flower. Pollination is effected as the insect struggles out of the body cavity. This interesting mode of pollination occurs also in *Caleana major*, the Flying Duck Orchid. Flowering December-January. Occasional in damp coastal heaths. Tas, Vic, NSW, Qld, SA; New Zealand.
Orchidaceae

Cyathodes juniperina *Pink Berry*
Usually a bushy shrub, resembling *Cyathodes parvifolia* but leaves longer, over 8 mm, tapering throughout the whole length to a long sharp point and often held at right angles to the stem. Branches frequently. Numerous small white bell-shaped flowers hanging down near the ends of branches. Fruits pink nearly 9 mm across, often scattered, but sometimes massed near the ends of branches. Sometimes larger shrub to 2 m or a small tree. Flowering November-December. Common on hillsides and on Central Plateau, and to sea level in many places in the north, west, south-west and Bass Strait islands. The only non-endemic species of *Cyathodes* in Tasmania, being found also in Victoria and New Zealand.
Epacridaceae

Leucopogon ericoides
Shrub to 1 m, bushy but also with long strong upright shoots, leaves linear-oblong, 1 cm long 2.5 mm wide, shortly stalked, margins strongly recurved, leaf apex abruptly pointed with a short hard point. Flowers in short dense axillary spikes along great lengths of stem (15-20 cm). Bracts and sepals finely hairy, sepals and buds pinkish, flowers white, tubular with spreading bearded lobes narrower than the tufts of hairs. Anthers near throat of the tube. Fruit a greenish-black drupe, scarcely succulent. A robust form common on the west coast in wet heaths has much longer wider leaves up to 4 mm wide. Flowering spring. Widespread, abundant in sandy and peaty heaths. Tas, Vic, NSW, Qld, SA.
Epacridaceae

Pultenaea hibbertioides
Small soft-leaved shrub with long spreading branches, almost prostrate or erect up to 2 m high; younger branches covered with short white hairs. Leaves linear 10-15 mm long, channelled above, dark green, with large conspicuous dark brown papery stipules, lower leaf surface rough. Terminal buds surrounded by wide brown bracts which remain on the stem for about one year after bud has grown on. Flowers orange-yellow, keel darker, in dense heads terminating short lateral branches surrounded by the clustered leaves. Flowering October-November. Very local in the north-east. Tas, Vic.
Fabaceae-Faboideae

Allocasuarina monilifera
Slender sheoak often only 50 cm high in coastal heaths but small tree on the Central Plateau. Male and female plants often separate. The leaves are scale-like teeth, in rings at the joints of the green slender twigs which are minutely ridged, the crest of the ridge running down from the base of each scale leaf, but no lines of white hairs visible. Male flowers in spikes at ends of branches, the distance between rings being less than on vegetative shoots. Young female cone stalked, red with long styles, mature cone grey, oblong, top often pointed, individual scales below each carpel quite conspicuous. Flowering summer. Common in heaths. Tas, Vic.
Casuarinaceae

268 *Hakea teretifolia* Dagger Hakea

269 *Paracaleana minor* Small Duck Orchid

270 *Cyathodes juniperina* Pink Berry

271 *Leucopogon ericoides*

272 *Pultenaea hibbertioides*

273 *Allocasuarina monilifera* (female)

274 *Allocasuarina monilifera* (male)

Melaleuca squarrosa *Scented Paper-bark*
A common erect shrub. The young shoots are often pubescent; old trunks have fawn papery bark, the roots and lower stems often knotted forming a hard woody persistent base (lignotuber) which survives fire, being buried in wet sandy soil, and gives rise to new upright shoots. Flower head, a short pale yellow bottle-brush spike, to 4 cm, having many sessile flowers with conspicuous stamens about 8 mm long. The stiff leaves 5-15 mm long are ovate, concave, slightly folded along the mid-rib, and pointed. Flowering October-December. Heaths, wet scrub and along watercourses, especially in north and east. Tas, Vic, NSW, SA.
Myrtaceae

Boronia anemonifolia *Stinking Boronia*
This boronia is variable in its leaf shape but characterised by its long stalked, sticky, very glandular leaves, smelling strongly of turpentine, and its numerous small flowers in groups shorter than the leaf below them. Leaves thick, warty and trifoliolate or each leaflet further lobed or divided, so appearing to have 5 or more leaflets. Leaves variable in width; a very wide-leafed form occurs on west coast. Flowers in stalked clusters; often 3 clusters on a common stalk and each cluster of up to 3 pink and white flowers. Plant in full flower is very showy with flowers in each leaf axil along many centimetres of stem. Flowering spring-summer. Scattered in heaths throughout the north and west, on dry hillsides in south. Tas, Vic, NSW, Qld.
Rutaceae

Ricinocarpos pinifolius *Wedding Bush*
Shrub to about 3 m with many large white flowers; separate male and female flowers on same plant. Leaves narrow, 2-3 cm long, 1-3 mm wide, upper surface hairless, margins recurved. Flowers stalked, 2-3 cm across, white, 4-6 petals. Male flowers more conspicuous than female, many stamens, joined in a central mass. Female flowers smaller, ovary covered with projections which become long, red and leathery in fruit. Ripe fruit spherical, 12 mm in diameter, bursting into 3 parts. Seeds mottled brown, 5 mm long with a fleshy white swollen knob at one end. Flowering spring-summer. North-east coastal heaths. Tas, Vic, NSW, Qld.
Euphorbiaceae

Dampiera stricta
A small slightly woody plant 20-40 cm high, sometimes compact, more often with several spreading angular or flattened stems, springing from a perennial base. Leaves rough-surfaced, hairy when young, 1-2 cm long, margins often toothed or lobed; upper leaves narrow, lower ones often irregularly wedge-shaped. Flowers pale to bright sky-blue, occasionally white, solitary or a few together in axils of upper leaves; back of petals covered with dense felt of brown hairs. Flowering September-October. Frequent on sandy soils near east and north-east coasts, and in Asbestos Range. The only Tasmanian representative of the endemic Australian genus *Dampiera* which contains over 50 species. Tas, Vic, NSW, Qld.
Goodeniaceae

Bossiaea cinerea
This pea with its brown and yellow flowers and narrow almost triangular leaves may be a mass of flowers in spring. An erect or spreading shrub with many long stiff branches 30-80 cm high. Leaves lanceolate 1-3 cm long tapering to a sharp point, with fine bristle-like stipules. Flowers 1 cm long, 1 or 2 together in axils along long lengths of stem, also on short lateral branches. Each flower long-stalked, calyx irregular, the two upper lobes broader than the three lower yellow with brown markings. Flattened pod about twice as long as broad. Flowering September-November. Abundant in coastal heaths, light coastal forest. Tas, Vic, NSW, SA.
Fabaceae-Faboideae

Hibbertia aspera
The long straggling branches of this erect shrub may sprawl over other plants and climb to a height of 2.5 m. Leaves elliptical-ovate or obovate, 1-2 cm long, rough textured, with very fine stellate hairs on the lower surface (use high magnification). Flowers 1-1.5 cm across, stalked, sepals silky hairy, petals obovate yellow, stamens grouped at one side of the hairy ovary; bark smooth brownish-grey. Flowering October-November. Known in a few isolated patches in north and north-east Tasmania, but may be more widespread. Tas, Furneaux Group, Vic.
Dilleniaceae

275 *Melaleuca squarrosa* Scented Paper-bark

276 *Boronia anemonifolia* Stinking Boronia

277 *Ricinocarpos pinifolius* Wedding Bush

278 *Ricinocarpos pinifolius* (fruits) Wedding Bush

279 *Dampiera stricta*

280 *Bossiaea cinerea*

281 *Hibbertia aspera*

Pimelea flava *Yellow Pimelea*

A shrub 0.5-1.5 m high with slender erect branches arising in whorls below the previous year's flower heads. Bark light brown, smooth, tearing in ribbons. Leaves 4-12 mm long, opposite, obovate-oblong or orbicular, blunt, bluish-green. Flower heads bright buttercup yellow, of many tubular flowers surrounded by four wide green bracts. Male and female flowers on different plants. Male flowers with wider spreading lobes, 2 orange stamens. Females a little narrower and less conspicuous. Fruit, a cluster of several hairy one-seeded fruits. Leaves turning bright blue-green when dried. Flowering September-October. Abundant but local near east and north coasts. Tas, Vic, NSW, SA.
Thymelaeaceae

Aotus ericoides *Golden Pea*

A small shrub to about 1 m branching from the base into long spreading stems crowded with axillary flowers. Flowers pea shaped, golden yellow with red markings, 2 or 3 in each leaf axil along the last 6-8 cm of stem making cylindrical heads. Leaves narrow 1-1.5 cm long, dark green, crowded, margins recurved to mid-rib; the short leaf stalks and bark on branches grey with soft velvety hairs. Pod small, ovate, velvety. Flowering October-November. Widespread and common especially in coastal areas. Tas, Vic, NSW, Qld, SA.
Fabaceae-Faboideae

Epacris barbata

A much-branched shrub to 1.2 m with large white flowers about 1 cm in diameter, the lobes slightly longer than the tube with the dark anthers slightly protruding. The floral bracts and sepals are tinged with red and covered with silky hairs. The leaves to 10 mm long are flat, elliptical, dark green and shining with a short stout stalk and a sharp point. Flowering spring. Found only in the vicinity of Freycinet Peninsula. Tas, endemic.
Epacridaceae

Xanthorrhoea australis *Grass Tree, Blackboy, Kangaroo Tail*

A perennial with an underground stem developing a tall trunk covered with rough densely packed leaf bases. Leaves long, harsh, narrow, grass-like, spirally arranged forming a hanging skirt around the trunk. Flowers in a dense hard columnar spike up to 1.5 m long, 10 cm diameter, on a stout stalk. Flowers very numerous, small, white, with perianth of 6 narrow strap-shaped parts, embedded in a mass of brown bracts. Fruits numerous sharp pointed capsules. Leaf bases and trunk exude a strong smelling resin which has been used as a substitute for shellac and is attractive to bees. Flowering in spring-summer or following fire. Foliage burns fiercely, leaving persistent crowded leaf bases. Widespread, coastal heaths and light forest. Tas, Vic, NSW, SA.
Xanthorrhoeaceae

Spyridium vexilliferum *Winged Spyridium*

Slender twiggy shrub, up to 90 cm high. Leaves distant and narrow-linear, blunt, mid-rib deeply impressed, margins recurved. Button-like flower heads surrounded by 1-3 velvety white floral wing-like bracts on slender stalks. Individual flowers white, with 5 sepals, 5 small petals forming hoods over stamens, nectar disc prominent. Flowering September-December. Locally common in sandy heaths, on river banks and rocky slopes in the east, north and west. Tas, SA, Vic, NSW.
Rhamnaceae

Patersonia fragilis *Short Purple Iris*

A small iris with flowers and tufts of leaves arising from a short creeping rhizome. Leaves thick, sometimes rounded, pale or grey-green, 15-30 cm long, firm, very narrow, pointed. Main flower stalks shorter than leaves, two long bracts enclosing flowers which open in succession. Flowers blue-mauve, 6-lobed, the 3 outer lobes large, rounded and spreading; the inner lobes tiny and erect. Fruit an angular capsule breaking into 3 segments and curling back still enclosed in the big brown bracts. There are two species of *Patersonia* in Tasmania; *P. fragilis* can be distinguished from *P. occidentalis* by the thicker leaves, short flower stalk and the tube of the flower which extends beyond the stem bracts. Flowering November-January. Common on sandy or gravelly soils near coasts, frequent in wet parts of coastal heaths. Tas, Vic, NSW, SA.
Iridaceae

282 *Pimelea flava* Yellow Pimelea

283 *Aotus ericoides* Golden Pea

284 *Epacris barbata*

285 *Xanthorrhoea australis*
Grass Tree, Blackboy, Kangaroo Tail

286 *Spyridium vexilliferum*
Winged Spyridium

287 *Patersonia fragilis*
Short Purple Iris

Patersonia occidentalis *Long-stalked Purple Iris*
This plant is similar to *Patersonia fragilis* in habit but can be distinguished, as the name suggests, by the length of the flowering stem which exceeds that of the thin flat leaves, being up to 40 cm. The tubes of the mauve flowers barely protrude beyond the bracts which loosely enfold them. Flowering November-January. Found in wet heathland habitats near the coast. Tas, Vic, NSW, SA.
Iridaceae

Baeckea ramosissima *Baeckea*
Small slender erect or prostrate shrub with long spreading branches. Leaves small, narrow-linear or oblong. Flowers variable in size 8-15 mm across, numerous, white or pinkish or white blotched with pink. Stamens 10. Fruit a capsule opening by 2-3 valves. Flowering spring-summer. Widespread, abundant in heaths and along rivers. Tas, Vic, NSW, SA.
Near Gladstone in north-east Tasmania a minute form occurs, prostrate with wiry stems, leaves 4 mm long, flowers about 5 mm across. It may be more widespread but overlooked.
Myrtaceae

Helichrysum dealbatum *White Everlasting*
A white paper daisy of wet lowland areas, perennial with small tufts of leaves springing from slender rhizomes. Flowering stems erect, unbranched and leafy. Leaves 2-4 cm long, linear-obovate, pointed, flat; upper surface dark dull green, silver-white on underside. Flower heads terminal, 2.5 cm in diameter, white, with reddish-purple flush on the outside bracts, eye of the daisy white. Flowering spring-summer. Common in wet peaty soils. Tas, Vic.
Asteraceae

Lyperanthus nigricans *Red Beaks, Mournful Flower*
Thick stemmed plant, 10-22 cm high with one fleshy heart-shaped spotted leaf about 5 cm long, flat on ground. Flowers 1-6, base of each enclosed in a green bract, the floral segments white streaked with reddish-purple, the uppermost forming a deep hood over the flower; labellum white and fringed. Blooms well after fires. Flowers turn black when dried, hence the name *nigricans.* Flowering spring. Widespread near coasts. Tas, all mainland States.
Orchidaceae

Cryptostylis subulata *Duckbill Orchid, Cow's Horns*
Fairly robust ground orchid with striking red and yellow flowers. Leaves 1-3, large, 6-8 cm long, stalked, elliptical, at base of plant. Flowering stem erect, 25-60 cm high, bearing 3-8 large stalked flowers with bracts below them. Sepals and petals narrow, yellowish, rolled inwards and spreading about base of the labellum. Labellum at upper side of flower ± horizontal, oblong, apex pointed, base yellow, concave downwards, then margins curving upward to expose the red underside with 2 dark purple ridges which end in thickened dark knobs near the apex. Column obscured in natural position. Flowering summer. Wet heaths, swampy flats. Tas, Vic, NSW, Qld.
Orchidaceae

Leptospermum scoparium *Manuka, Tea-tree*
Common shrub, sometimes a tree with spreading or erect branches, leaves elliptical-lanceolate, pointed, concave above, stiff. Flowers white, solitary in leaf axils, with 5 small triangular sepals, 5 white round petals, many stamens surrounding a flat nectar disc, style short. Fruit, capsule with hemispherical base and rounded above, opening by 5 slits. Seeds small, fine, like short curved hairs. Flowering spring-summer. Very common, widespread in heaths, becoming a small tree to 6 m in wet forests. Tas, Vic, NSW, Qld; New Zealand.
Myrtaceae

288 *Patersonia occidentalis*
Long-stalked Purple Iris

289 *Baeckea ramosissima*
Baeckea

290 *Helichrysum dealbatum*
White Everlasting

291 *Lyperanthus nigricans*
Red Beaks, Mournful Flower

292 *Cryptostylis subulata*
Duckbill Orchid, Cow's Horns

293 *Leptospermum scoparium*
Manuka, Tea-tree

Leptospermum scoparium var. **eximium**
A densely branched shrub to 2 m. Leaves wider than in other forms of *L. scoparium*, broadly ovate to almost orbicular, pointed, shining. Flowers numerous, white, 1.5-2 cm across, terminal on short axillary branches. Fruit like that of *L. scoparium*, base hemispherical, domed on top. Flowering November-December. Local in south and south-west. Tas, endemic.
Myrtaceae

Boronia nana
Small plant with a woody underground stem and several thin prostrate branches about 15 cm long. Leaves linear, pointed, flat, thick, occasionally trifoliolate with narrow leaflets. Small axillary flowers with 4 ovate sepals, 4 pink petals, 3-5 mm long, twice as long as sepals, 8 stamens finely hairy in the lower half. Fruit formed from 4 carpels, separating into 4 parts. Flowering October-December. Sandy heaths and other places near the north coast. Tas, Vic.
Rutaceae

Lagurus ovatus *Hare's Tail Grass*
A small annual grass with pale or purplish heads up to 45 cm high, and flat grey green leaves covered with soft hairs. Flowering head an attractive soft ovoid head of many spikelets surrounded by soft straight hairs. Each tiny spikelet has two outer bracts covered with soft hairs, bract tips are drawn out into plumose bristles. Inner bract against the seed has two short bristles and a longer soft awn; these form the long outstanding hairs of the head. Common in sandy soil especially those containing lime. Said to indicate copper deficiency. Sometimes grown as a border plant. Common near coasts.
Poaceae

Nablonium calyceroides
A small perennial daisy with a basal rosette of leaves 4-5 cm long, white underneath; the wide stalks about the same length as the lanceolate blades. Flower heads white, solitary, terminal on stout erect stems 4-8 cm long; outer bracts greenish with papery margins, no ray florets, tube florets white. Seeds flattened, ridged, with two large divergent spines at the apex; spines hard and sharp. Flowering December. Wet flats behind dunes, margins of lagoons on Bass Strait islands and a few localities on the west coast. Tas, endemic.
Asteraceae

Epacris myrtifolia
A robust bushy shrub 30-60 cm high. Leaves crowded, hard, often ribbed underneath, elliptical-ovate, rather wide, blade flat or concave. Flowers white, 5 mm long, solitary in the leaf axils and crowded near branch ends in dense showy heads. Anthers at throat of tube, style not protruding. Flowering October-December. Local near coasts in the south. Tas, endemic.
Epacridaceae

Caladenia caudata
A rather short spider orchid, the flower often in shades of crimson and purple, sometimes yellow-green with red markings. One linear or lanceolate hairy leaf sheathes the base of an erect stem 6-15 cm long. Flowers terminal, 1 or 2 together, dorsal sepal erect, sepals and petals linear, spreading then curving down in long drawn out glandular tails. Differs from *C. patersonii* in the labellum (tongue) which arches then curves down into a long narrow tail-like point; margins with long fine teeth decreasing to serrations at the tip. Labellum with 4-6 rows of erect dark coloured glands (calli) from base to just beyond the bend. Flowering October. Local on warm hillsides near the coast in the west, north and south-east. Tas, endemic.
Orchidaceae

Hakea sericea *Silky Hakea*
Much branched shrub with sharp needle-shaped leaves, up to 4 m tall, young branches with red-brown bark. Leaves cylindrical, slender with long pungent points, 3-9 cm long. Flowers white or pink, tube hairless, slender, 5-6 mm long, 2-6 flowers together in clusters in the upper leaf axils making long leafy sprays. Fruit, brown woody capsule, 25-30 mm long, 17-22 mm wide, outer surface smooth or ridged with two prominent upturned points. Seeds winged, their inner surface rough with minute projections. Flowers early spring. Tas (Flinders and Cape Barren islands), Vic, NSW. A cultivated variety with pink and white flowers is available.
Proteaceae

294 *Leptospermum scoparium* var. *eximium*

295 *Boronia nana*

296 *Lagurus ovata* Hare's Tail Grass

297 *Nablonium calyceroides*

298 *Epacris myrtifolia*

299 *Caladenia caudata*

300 *Hakea sericea* Silky Hakea

INDEX

Bold figures indicate colour plate numbers